A FIRST COURSE IN INTEGRAL EQUATIONS

A FIRST COURSE IN INTEGRAL EQUATIONS

Abdul-Majid Wazwaz
Saint Xavier University, USA

Published by

World Scientific Publishing Co. Pte. Ltd.
P O Box 128, Farrer Road, Singapore 912805
USA office: Suite 1B, 1060 Main Street, River Edge, NJ 07661
UK office: 57 Shelton Street, Covent Garden, London WC2H 9HE

Library of Congress Cataloging-in-Publication Data
Wazwaz, Abdul-Majid.
A first course in integral equations / Abdul-Majid Wazwaz.
p. cm.
Includes bibliographical references and index.
ISBN 9810231016
1. Integral equations. I. Title.
QA431.W36 1997
515'.45--dc21 97-3749
CIP

British Library Cataloguing-in-Publication Data
A catalogue record for this book is available from the British Library.

This book is printed on acid-free paper.

Printed in Singapore by Uto-Print

Contents

1 Introductory Concepts **1**
1.1 Definitions 1
1.2 Classification of Linear Integral Equations 3
1.2.1 Fredholm Linear Integral Equations: 3
1.2.2 Volterra Linear Integral Equations: 4
1.2.3 Integro–Differential Equations: 6
1.2.4 Singular Integral Equations: 7
1.3 Solution of an Integral Equation 11
1.4 Converting Volterra Equation to ODE 15
1.5 Converting IVP to Volterra Equation 20
1.6 Converting BVP to Fredholm Equation 26

2 Fredholm Integral Equations **31**
2.1 Introduction 31
2.2 The Decomposition Method 33
2.2.1 The Modified Decomposition Method: 38
2.3 The Direct Computation Method 43
2.4 The Successive Approximations Method 48
2.5 The Method of Successive Substitutions 52
2.6 Comparison between Alternative Methods 56
2.7 Homogeneous Fredholm Equations 59

3 Volterra Integral Equations **67**
3.1 Introduction 67
3.2 The Adomian Decomposition Method 68
3.2.1 The Modified Decomposition Method: 73
3.3 The Series Solution Method 77

3.4 Converting Volterra Equation to IVP 82
3.5 Successive Approximations Method 86
3.6 The Method of Successive Substitutions 91
3.7 Comparison between Alternative Methods 95
3.8 Volterra Equations of the First Kind 99

4 Integro-Differential Equations 103
4.1 Introduction 103
4.2 Fredholm Integro-Differential Equations 105
4.2.1 The Direct Computation Method: 105
4.2.2 The Adomian Decomposition Method: 109
4.2.3 Converting to Fredholm Integral Equations: . . . 118
4.3 Volterra Integro-Differential Equations 121
4.3.1 The Series Solution Method: 121
4.3.2 The Decomposition Method: 126
4.3.3 Converting to Volterra Integral Equation: 131
4.3.4 Converting to Initial Value Problems: 134

5 Singular Integral Equations 139
5.1 Definitions 139
5.2 Abel's Problem 141
5.2.1 The Generalized Abel's Integral Equation 146
5.3 The Weakly-Singular Volterra Equations 150

6 Nonlinear Integral Equations 157
6.1 Definitions 157
6.2 Nonlinear Fredholm Integral Equations 158
6.2.1 The Direct Computation Method 159
6.2.2 The Decomposition Method 163
6.3 Nonlinear Volterra Integral Equations 173
6.3.1 The Series Solution Method 173
6.3.2 The Decomposition Method 177

A Table of Integrals 183

B Integrals of Irrational Functions 187

C Series Representations 189

D The Error and Gamma Functions **191**

Answers To Exercises **193**

Bibliography **205**

Index **207**

Preface

Mathematics, science and engineering students, both advanced undergraduate and beginning graduate, need an integral equations textbook that simply and easily introduces the material. They also need a textbook that embarks upon their already acquired knowledge of regular integral calculus and ordinary differential equations. Because of these needs, this textbook was created. From many years of teaching, I have found that the available treatments of the subject are abstract. Moreover, most of them are based on comprehensive theories such as topological methods of functional analysis, Lebesgue integrals and Green functions. Such methods of introduction are not easily accessible to those who have not yet had a background in advanced mathematical concepts. This book is especially designed for those who wish to understand integral equations without having the extensive mathematical background. In this fashion, this text leaves out abstract methods, comprehensive methods and advanced mathematical topics.

From my experience in teaching and in guiding related senior seminar projects for advanced undergraduate students, I have found that the material can indeed be taught in an accessible manner. Students have shown both a lot of motivation and cabability to grasp the subject once the abstract theories were ommitted. In my approach to teaching integral equations, I focus on easily applicable techniques and I don't emphasize such abstract methods as existence, uniqueness, convergence and Green functions. I have translated my means of introducing and fully teaching this subject into this text so that the intended user can take full advantage of the easily presented and explained material.

I have also introduced and made full use of some recent developments in this field.

The book consists of six chapters, each being divided into sections. In each chapter the equations are numbered consecutively and distinctly from other chapters. Several examples are introduced in each section, and a large number of exercises, with varying degrees of difficulty but being consistent with the material, are included to give the students constructive insights about the material and to provide them with useful practice.

In this text, we were mainly concerned with linear integral equations, mostly of the second kind. Chapter 1 introduces classifictions of integral equations and necessary techniques to convert differential equations to integral equations or vice versa. Chapter 2 deals with linear Fredholm integral equations and the reliable techniques, supported by the new developments, to handle this style of equations. In Chapter 3 the linear Volterra integral equations are handled, using the recent developed techniques beside standard ones. In Chapter 4 the topic of integro-differential equations has been handled and reliable techniques were implemented to handle the essential link between differential and integral operators. Chapter 5 introduces the treatment of the singular and the weakly singular Volterra type integral equations. Chapter 6 deals with the nonlinear integral equations. This topic is difficult to study. However, recent developments have shown improvements over existing techniques and allow this topic to be far more easily accessible for specific cases. A large number of nonlinear integral examples and exercises are investigated.

Throughout the text, examples are provided to clearly and throughly introduce the new material in a clear and absorbable fashion. Many exercises are provided to give the new learner a chance to bulid his confidence, ease and skill with the newly learned material.

The text has four useful Appendices. These Appendices provide the user with the integral forms, Maclaurin series and other related materials which are needed to be used in the exercises.

Finally, this book is suitable for a one-semester course in integral equations, hoping it will be useful for anyone intersted in integral equations.

I am indebted to my wife, my son and my daughters who provided me with their encouragement, patience and support.

I am also indebted to Professor Louis Pennisi who first introduced me to integral equations and instilled in me great love for it. His excellent methods of teaching have inspired me greatly.

The author would highly appreciate any note concerning any error found and for any constructive suggestion.

Chicago, IL **1996** Abdul-Majid Wazwaz

Chapter 1

Introductory Concepts

1.1 Definitions

An integral equation is an equation in which the unknown function u(x) to be determined appears under the integral sign. A typical form of an integral equation in u(x) is of the form

$$u(x) = f(x) + \int_{\alpha(x)}^{\beta(x)} K(x,t)\, u(t)\, dt, \tag{1}$$

where K(x,t) is called the kernel of the integral equation, and $\alpha(x)$ and $\beta(x)$ are the limits of integration. In (1), it is easily observed that the unknown function u(x) appears under the integral sign as stated above, and out of the integral sign in most other cases. It is important to point out that the kernel K(x,t) and the function f(x) in (1) are given in advance. Our goal is to determine u(x) that will satisfy (1), and this may be achieved by using different techniques that will be discussed in the forthcoming chapters. The primary concern of this text will be focused on introducing these methods and techniques supported by illustrative and practical examples.

Integral equations arise naturally in physics, chemistry, biology and engineering applications modelled by initial value problems for a finite interval $[a, b]$. More details about the sources and origins of integral equations can be found in [12] and [14]. In the following example we will discuss how an initial value problem will be converted to the form of an integral equation.

Example 1. Consider the initial value problem

$$u'(x) = 2xu(x), \quad x \geq 0, \tag{2}$$

subject to the initial condition

$$u(0) = 1. \tag{3}$$

The equation (2) can be easily solved by using separation of variables; where by using the initial condition (3), the solution

$$u(x) = e^{x^2}, \tag{4}$$

is easily obtained. However, integrating both sides of (2) with respect to x from 0 to x and using the initial condition (3) yield the following

$$\int_0^x u'(t)\,dt = \int_0^x 2tu(t)dt, \tag{5}$$

or equivalently

$$u(x) = 1 + \int_0^x 2tu(t)\,dt, \tag{6}$$

obtained by integrating the left hand side of (5) and by using the given initial condition (3). Comparing (6) with (1) we find that f(x)=1 and the kernel K(x,t) =2t.

We will further discuss the algorithms of converting initial value problems and boundary value problems in detail to equivalent integral equations in the forthcoming sections. As stated above, our task is to determine the unknown function u(x) that appears under the integral sign as in (1) and (6) and that will satisfy the given integral equation.

We further point out that integral equations as (1) and (6) are called *linear* integral equations. This classification is used if the unknown function u(x) under the integral sign occurs linearly i.e. to the first power. However, if u(x) under the integral sign is replaced by a nonlinear function in u(x), such as $u^2(x)$, $\cos u(x)$ and $e^{u(x)}$, *etc.*, the integral equation is called in this case a *nonlinear* integral equation.

1.2 Classification of Linear Integral Equations

The most frequently used linear integral equations fall under two main classes namely Fredholm and Volterra integral equations. However, in this text we will distinguish four types of linear integral equations; the two main classes and two related types of integral equations. In particular, the four types are given by:

1. **Fredholm integral equations**

2. **Volterra integral equations**

3. **Integro-Differential equations**

4. **Singular integral equations.**

In the following we will outline the basic definitions and properties of each type.

1.2.1 Fredholm Linear Integral Equations:

The standard form of Fredholm linear integral equations, where the limits of integration a and b are constants, are given by the form

$$\phi(x)\,u(x) = f(x) + \lambda \int_a^b K(x,t)u(t)\,dt,\; a \leq x,\; t \leq b, \tag{7}$$

where the kernel of the integral equation K(x,t) and the function f(x) are given in advance, and λ is a parameter. The equation (7) is called *linear* because the unknown function $u(x)$ under the integral sign occurs linearly, i.e. the power of $u(x)$ is one.

The value of ϕ (x) will give the following kinds of Fredholm linear integral equations:

1. When ϕ (x) = 0, Eq. (7) becomes

$$f(x) + \lambda \int_a^b K(x,t)u(t)\,dt = 0, \tag{8}$$

and is called Fredholm integral equation of the first kind.

2. When ϕ(x) =1, Eq. (7) becomes

$$u(x) = f(x) + \lambda \int_a^b K(x,t)u(t)dt, \tag{9}$$

and is called Fredholm integral equation of the second kind. In fact, the equation (9) can be obtained from (7) by dividing both sides of (7) by ϕ (x) provided that ϕ (x) $\neq$ 0.

1.2.2 Volterra Linear Integral Equations:

The standard form of Volterra linear integral equations, where the limits of integration are functions of x rather than constants, are of the form

$$\phi(x)\,u(x) = f(x) + \lambda \int_a^x K(x,t)u(t)\,dt, \tag{10}$$

where the unknown function $u(x)$ under the integral sign occurs linearly as stated before. It is worth noting that (10) can be viewed as a special case of Fredholm integral equation when the kernel K(x,t) vanishes for $t > x$, x is in the range of integration $[a, b]$.

As in Fredholm equations, Volterra integral equations fall under two kinds, depending on the value of ϕ(x), namely:

1. When ϕ (x) = 0, Eq. (10) becomes

$$f(x) + \lambda \int_a^x K(x,t)u(t)\,dt = 0, \tag{11}$$

and is called Volterra integral equation of the first kind.

2. When ϕ(x) =1, Eq. (10) becomes

$$u(x) = f(x) + \lambda \int_a^x K(x,t)u(t)dt, \tag{12}$$

and is called Volterra integral equation of the second kind.

Examining the equations (7) – (12) carefully, the following remarks can be concluded:

1. **The structure of Fredholm and Volterra equations:**

   The unknown function $u(x)$ appears linearly only under the integral sign in linear Fredholm and Volterra integral equations of the first kind. However, the unknown function $u(x)$ appears linearly under the integral sign and out of the integral sign as well in the second kind of these linear integral equations.

2. **The limits of integration:**

   In Fredholm integral equations, the integral is taken over a finite interval with fixed limits of integration. However, in Volterra integral equations, at least one limit of the range of integration is a variable, and the upper limit is the most commonly used with a variable limit.

3. **The origins of integral equations:**

   It is important to note that integral equations arise in engineering, physics, chemistry, and biology problems [12] and [14]. Further, integral equations arise as representation forms of differential equations. Furthermore, Fredholm and Volterra integral equations arise from different origins and applications, such as boundary value problems as in Fredholm equations, and from initial value problems as in Volterra equations. Based on the fact that integral equations arise from distinct origins, different techniques and approaches will be used to determine the solution of each type of integral equations.

4. **The linearity property:**

   As indicated before, the unknown function u(x) in Fredholm and Volterra integral equations (9) and (12) occurs to the first power under the integral sign. However, nonlinear Fredholm and Volterra integral equations arise if u(x) is replaced by a nonlinear function F(u(x)). The following are examples of nonlinear integral equations:

$$u(x) = f(x) + \lambda \int_a^x K(x,t)u^2(t)dt, \qquad (13)$$

$$u(x) = f(x) + \lambda \int_a^x K(x,t)e^{u(t)}dt, \qquad (14)$$

$$u(x) = f(x) + \lambda \int_0^1 K(x,t)\sin(u(t))dt, \tag{15}$$

where the linear function u(x) in (1) has been replaced by the nonlinear functions $u^2(t)$, $e^{u(t)}$ and sin(u(t)) respectively.

5. **The homogeneity property:**

   If we set f(x) = 0 in Fredholm or Volterra integral equation of the second kind given by (9) and (12), the resulting equation is called a *homogeneous* integral equation, otherwise it is called *nonhomogeneous* integral equation.

6. **The singular behavior of the integral equation:**

   An integral equation is called singular if the integration is improper. This usually occurs if the interval of integration is *infinite*, or if the kernel becomes *unbounded* at one or more points of the interval of consideration a $\leq$ t $\leq$ b. Singular integral equations will be defined soon, and the methods to handle it will be introduced later in Chapter 5.

It is worth noting that two other types of integral equations, related to the two main classes Fredholm and Volterra integral equations arise in many science and engineering applications. In the following, we introduce these significant equations as distinct types.

1.2.3 Integro–Differential Equations:

Volterra, in the early 1900, studied the population growth, where new type of equations have been developed and was termed as integro–differential equations. In this type of equations, the unknown function u(x) occurs in one side as an *ordinary derivative*, and appears on the other side under the *integral sign*. Several phenomenas in physics and biology [14] and [20] give rise to this type of integro–differential equations. Further, we point out that an integro–differential equation can be easily observed as an intermediate stage when we convert a differential equation to an integral equation as will be discussed later in the coming sections.

The following are examples of integro–differential equations:

$$u''(x) = -x + \int_0^x (x-t)u(t)dt, \quad u(0) = 0, \ u'(0) = 1, \tag{16}$$

$$u'(x) = -\sin x - 1 - + \int_0^x u(t)dt, \quad u(0) = 1, \tag{17}$$

$$u'(x) = 1 - \frac{1}{3}x + \int_0^1 xtu(t)dt, \quad u(0) = 1. \tag{18}$$

Equations (16) and (17) are called integro–differential equations related to Volterra integral equations, or Volterra *integro-differential* equations. However, equation (18) is called integro–differential equation related to Fredholm integral equations, or simply Fredholm *integro-differential* equation . This classification has been concluded as a result to the limits of integration. The solution for integro–differential equations will be established using in particular the most recent developed techniques. The integro-differential equations will be discussed extensively in Chapter 4.

1.2.4 Singular Integral Equations:

The integral equation of the first kind

$$f(x) = \lambda \int_{\alpha(x)}^{\beta(x)} K(x,t)\, u(t)dt. \tag{19}$$

or the integral equation of the second kind

$$u(x) = f(x) + \lambda \int_{\alpha(x)}^{\beta(x)} K(x,t)\, u(t)dt, \tag{20}$$

is called *singular* if the lower limit, the upper limit or both limits of integration are *infinite*. In addition, the equations (19) or (20) is also called a singular integral equation if the kernel $K(x,t)$ becomes *infinite* at one or more points in the domain of integration. Examples of the first type of *singular* integral equations are given by the following examples:

$$u(x) = 2x + 6\int_0^\infty \sin(x-t)u(t)dt, \tag{21}$$

$$u(x) = x + \frac{1}{3}\int_{-\infty}^{0} \cos(x+t)u(t)dt, \tag{22}$$

$$u(x) = 1 + x^2 + \frac{1}{6}\int_{-\infty}^{\infty} (x+t)u(t)dt, \tag{23}$$

where the singular behavior in these examples has resulted from the range of integration becoming *infinite.*

Examples of the second kind of *singular* integral equations are given by

$$x^2 = \int_0^x \frac{1}{\sqrt{x-t}}u(t)dt, \tag{24}$$

$$x = \int_0^x \frac{1}{(x-t)^\alpha}u(t)dt, \quad 0 < \alpha < 1, \tag{25}$$

$$u(x) = 1 - 2\sqrt{x} - \int_0^x \frac{1}{\sqrt{x-t}}u(t)dt, \tag{26}$$

where the singular behavior in this kind of equations has resulted from the kernel K(x,t) becoming *infinite* as $t \to x$.

It is important to note that integral equations similar to examples (24) and (25) are called Abel's integral equation and generalized Abel's integral equation respectively. Moreover these types of singular integral equations are among the earliest integral equations established by the Norwegian mathematician Niels Abel in 1823. Singular equations similar to example (26) are called the weakly-singular second-kind Volterra type integral equations. This type of equations usually arise in science and engineering applications like heat conduction, superfluidity and crystal growth. The singular integral equations and the methods to handle it will be discussed in Chapter 5.

In closing this section, we illustrate the classifications and the basic concepts that were discussed above by the following examples.

Example 2. Classify the following integral equation

$$u(x) = x - \frac{1}{6}x^3 + \int_0^x (x-t)u(t)dt, \tag{27}$$

as Fredholm or Volterra integral equation, *linear* or *nonlinear* and *homogeneous* or *nonhomogeneous.*

Noting that the upper limit of the integral is x indicates that the equation (27) is a Volterra integral equation of the second kind. The equation (27) is *linear* since the unknown function $u(t)$ appears linearly under the integral sign. The presence of the function $f(x) = x - \frac{1}{3}x^3$ classifies the equation as a *nonhomogeneous* equation.

Example 3. Classify the following integral equation

$$u(x) = \frac{1}{2} + x - \int_0^1 (x-t)u^2(t)dt, \tag{28}$$

as Fredholm or Volterra integral equation, *linear* or *nonlinear* and *homogeneous* or *nonhomogeneous.*

The limits of integration are constants, therefore the equation (28) is a Fredholm integral equation. Further, the unknown function appears under the integral sign with power two indicating the equation is a *nonlinear* equation. The nonhomogeneous part f(x) appears in the equation showing that it is a *nonhomogeneous* equation.

Finally we discuss the following example.

Example 4. Classify the following equation

$$u'(x) = 1 - \frac{1}{3}x^3 + \int_0^x tu(t)dt, \quad u(0) = 0, \tag{29}$$

as Fredholm or Volterra integro-differential equation, and *linear* or *nonlinear.*

It is easily seen that (29) includes differential and integral operators, and by noting that the upper limit of the integral is a variable, we conclude that (29) is a Volterra integro-differential equation. Moreover, the equation is *linear* since $u(x)$ and $u'(x)$ appear linearly in the equation.

We point out that linear Fredholm integral equations, linear Volterra integral equations, Integro–differential equations and Singular integral equations will be discussed in Chapters 2, 3, 4 and 5 respectively. The nonlinear integral equations will be discussed briefly in Chapter 6. The recent developed methods, that proved its effectiveness and reliability, will be applied to all types of integral equations.

Exercises 1.2

In exercises 1 – 10, classify each of the following integral equations as Fredholm or Volterra integral equation, *linear* or *nonlinear*, and *homogeneous* or *nonhomogeneous*:

(1) $u(x) = x + \int_0^1 xtu(t)\,dt$

(2) $u(x) = 1 + x^2 + \int_0^x (x-t)u(t)\,dt$

(3) $u(x) = e^x + \int_0^x tu^2(t)\,dt$

(4) $u(x) = \int_0^1 (x-t)^2\,u(t)\,dt$

(5) $u(x) = \frac{2}{3}x + \int_0^1 xtu(t)\,dt$

(6) $u(x) = \frac{3}{4}x + \frac{1}{5} + \int_0^1 (x-t)^3\,u(t)\,dt$

(7) $u(x) = 1 + \frac{x}{4}\int_0^1 \frac{1}{x+t}\,\frac{1}{u(t)}\,dt$

(8) $u(x) = \frac{1}{2}\cos x + \frac{1}{2}\int_0^{\frac{\pi}{2}} \cos x\,u(t)\,dt$

(9) $u(x) = 1 + \int_0^x (x-t)^2\,u^2(t)\,dt$

(10) $u(x) = 1 - \int_0^x (x-t)\,u(t)\,dt$

In exercises 11 – 15, classify each of the following integro-differential equations as Fredholm integro-differential equation or Volterra integro-differential equation. Also determine whether the equation is *linear* or *nonlinear*.

(11) $u'(x) = 1 + \int_0^x e^{-2t}u^3(t)\,dt, \quad u(0) = 1$

(12) $u'(x) = 1 - \frac{1}{3}x + \int_0^1 xtu(t)\,dt, \quad u(0) = 0$

(13) $u''(x) = \frac{1}{2}x^2 - \int_0^x (x-t)u^3(t)\,dt, \quad u(0) = 1, u'(0) = 0$

(14) $u'''(x) = \sin x - x + \int_0^{\frac{\pi}{2}} xtu'(t)\,dt,$

$$u(0) = 1, u'(0) = 0, u''(0) = -1$$

(15) $u'''(x) = -\frac{1}{12}x^4 + \int_0^x (x-t)u(t)\,dt,$

$$u(0) = u'(0) = 0, u''(0) = 2$$

In exercises 16 – 20, integrate both sides of each of the following differential equations once from 0 to x, and use the given initial condition to convert to a corresponding integral equation or integro-differential equation. (Follow example 1)

(16) $u'(x) = 4u(x), \quad u(0) = 1.$

(17) $u'(x) = 3x^2u(x), \quad u(0) = 1.$

(18) $u'(x) = u^2(x), \quad u(0) = 4.$

(19) $u''(x) = 4xu^2(x), \quad u(0) = 2, u'(0) = 1.$

(20) $u''(x) = 2xu(x), \quad u(0) = 0, u'(0) = 1.$

1.3 Solution of an Integral Equation

A solution of an integral equation or an integro-differential equation on the interval of integration is a function u(x) such that it satisfies the given equation. In other words, if the given solution is substituted in the right hand side of the equation, the output of this direct substitution must yield the left hand side, i.e. we should verify that the given function $u(x)$ satisfies the integral equation or the integro-differential equation under discussion. This important concept will be illustrated first by examining the following examples.

Example 1. Show that u(x) = e^x is a solution of the Volterra integral equation

$$u(x) = 1 + \int_0^x u(t)dt. \tag{30}$$

Substituting u(x) = e^x in the right hand side (RHS) of (30) yields

$$\begin{aligned} RHS &= 1 + \int_0^x e^t dt, \\ &= 1 + [e^t]_0^x \\ &= e^x, \\ &= u(x), \\ &= LHS. \end{aligned} \tag{31}$$

Example 2. Show that u(x) = x is a solution of the following Fredholm integral equation

$$u(x) = \frac{5}{6}x - \frac{1}{9} + \frac{1}{3}\int_0^1 (x+t)u(t)dt. \tag{32}$$

Substituting u(x) = x in the right hand side of (32) we obtain

$$\begin{aligned} RHS &= \frac{5}{6}x - \frac{1}{9} + \frac{1}{3}\int_0^1 (x+t)u(t)dt, \\ &= \tfrac{5}{6}x - \tfrac{1}{9} + \tfrac{1}{3}[\frac{xt^2}{2} + \frac{t^3}{3}]_0^1, \\ &= x \\ &= u(x), \\ &= LHS. \end{aligned} \tag{33}$$

In the following example we will discuss the solution of an integro-differential equation.

Example 3. Show that u(x) = x is a solution of the following Fredholm integro–differential equation

$$u'(x) = \frac{2}{3} + \int_0^1 tu(t)dt. \tag{34}$$

Substituting u(x) = x in the right hand side of (34) we obtain

$$\begin{aligned} RHS &= \frac{2}{3} + \int_0^1 t^2\,dt, \\ &= \frac{2}{3} + [\frac{t^3}{3}]_0^1 \\ &= 1 \\ &= u'(x) \\ &= LHS. \end{aligned} \tag{35}$$

Three valuable remarks can be made with respect to the concept of the solution of an integral equation or an integro-differential equation. First, the question of *existence* of a solution and the question of *uniqueness* of a solution, that usually we discuss in differential equations and integral equations will be left for further studies.

We next remark that if a solution exists for an integral equation or an integro-differential equation, it is important to note that this solution may be given in a closed form expressed in terms of elementary functions, such as a polynomial, exponential,trigonometric or hyperbolic function, similar to the solutions given in examples $1-3$. However, it is not always possible to obtain the solution in a closed form, but instead the solution obtained may be expressible in a series form. The solution obtained in a series form is usually used for numerical approximations, and in this case the more terms obtained will result in a better accuracy level.

It is important to illustrate the difference between the two expressible forms, the exact solution in a closed form and the approximant solution in a series form. Considering Example 1 above, we note that the exact solution is given in a closed form by the exponential function

$u(x) = e^x$. However, it will be shown later that the solution of the integral equation

$$u(x) = 1 + \frac{1}{4}\int_0^x xu(t)dt, \tag{36}$$

is given by the series form

$$u(x) = 1 + \frac{1}{4}x^2 + \frac{1}{48}x^4 + \frac{1}{960}x^6 + \cdots, \tag{37}$$

where we can easily observe that it is difficult to express the series (37) in an equivalent closed form. As indicated earlier, the series obtained can be employed to provide numerical approximations, and to achieve the highly desirable accuracy we should determine more terms in the series solution.

In closing our remarks, we consider the nonlinear integral equation

$$u(x) = \frac{5}{6}x + \frac{1}{2}\int_0^1 xu^2(t)dt. \tag{38}$$

It was found that equation (38) has two real solutions given by

$$u(x) = x,\ 5x, \tag{39}$$

and this can be justified through direct substitution. The *uniqueness* concept is not applicable for this example and for many other nonlinear problems. The nonlinear problems will be examined briefly in Chapter 6. The uniqueness criteria for nonlinear problems is justified for only specific problems under specific conditions. Generally speaking nonlinear integral equations are difficult to handle.

It is useful to point out that our main concern in this text will be on the linear integral equations and the linear integro-differential equations only. In addition, we will focus our study on equations with closed form solutions as in Examples 1 and 2. Other cases that may lead to a series solution will be investigated as well, supported by the development of the reliable techniques that will be discussed later. Moreover, nonlinear integral equations will be investigated in its simplest forms in Chapter 6. The recent developed methods presented powerful techniques, and therefore these methods have been carried out with promising results in linear and nonlinear equations.

Exercises 1.3

In exercises 1 – 10, verify that the given function is a solution of the corresponding integral or integro-differential equation :

(1) $u(x) = x + \int_0^{\frac{1}{4}} u(t)dt, \quad u(x) = x + \frac{1}{24}$

(2) $u(x) = \frac{2}{3}x + \int_0^1 xtu(t)dt, \quad u(x) = x$

(3) $u(x) = x + \int_0^1 xtu^2(t)dt, \quad u(x) = 2x$

(4) $u(x) = x - \int_0^x (x-t)u(t)dt, \quad u(x) = \sin x$

(5) $u(x) = 2\cosh x - x\sinh x - 1 + \int_0^x tu(t)dt, \quad u(x) = \cosh x$

(6) $u(x) = x + \frac{1}{5}x^5 - \int_0^x tu^3(t)dt, \quad u(x) = x$

(7) $u'(x) = 2x - x^4 + \int_0^x 4tu(t)dt, \ u(0) = 0, \quad u(x) = x^2$

(8) $u''(x) = x\cos x - 2\sin x + \int_0^x tu(t)dt,$

$$u(0) = 0, \ u'(0) = 1, \quad u(x) = \sin x$$

(9) $\int_0^x (x-t)^2 u(t)dt = x^3, \quad u(x) = 3$

(10) $\int_0^x (x-t)^{1/2} u(t)dt = x^{3/2}, \quad u(x) = \frac{3}{2}$

1.4 Converting Volterra Equation to ODE

In this section we will present the technique that converts Volterra integral equations of the second kind to equivalent differential equations. This may be easily achieved by applying the important Leibnitz Rule for differentiating an integral. It seems reasonable to review the basic outline of the rule.

To differentiate the integral $\int_{\alpha(x)}^{\beta(x)} G(x,t)dt$ with respect to x, we usually apply the useful Leibnitz rule given by:

$$\frac{d}{dx}\int_{\alpha(x)}^{\beta(x)} G(x,t)dt = G(x,\beta(x))\frac{d\beta}{dx} - G(x,\alpha(x))\frac{d\alpha}{dx} + \int_{\alpha(x)}^{\beta(x)}\frac{\partial G}{\partial x}\,dt, \tag{40}$$

where G(x,t) and $\frac{\partial G}{\partial x}$ are continuous functions in the domain D in the xt-plane that contains the rectangular region R, $a \le x \le b$, $t_0 \le t \le t_1$, and the limits of integration $\alpha(x)$ and $\beta(x)$ are defined functions having continuous derivatives for $a < x < b$. We note that Leibnitz rule is usually presented in most calculus books, and our concern will be on using the rule rather than its theoretical proof. The following examples are illustrative and will be mostly used in the coming approach that will be used to convert Volterra integral equations to differential equations.

Example 1. Find $\displaystyle\frac{d}{dx}\int_0^x (x-t)^2 u(t)\,dt.$

In this example, $\alpha(x) = 0$, $\beta(x) = x$, hence $\alpha'(x) = 0$, $\beta'(x) = 1$ and $\displaystyle\frac{\partial G}{\partial x} = 2(x-t)u(t)$. Using Leibnitz rule (40), we find

$$\frac{d}{dx}\int_0^x (x-t)^2 u(t)\,dt = \int_0^x 2(x-t)u(t)\,dt. \tag{41}$$

Example 2. Find $\displaystyle\frac{d}{dx}\int_0^x (x-t)u(t)\,dt.$

In this example, $\alpha(x) = 0$, $\beta(x) = x$, hence $\alpha'(x) = 0$, $\beta'(x) = 1$, and $\frac{\partial G}{\partial x} = u(t)$. Using Leibnitz rule (40), we find

$$\frac{d}{dx}\int_0^x (x-t)u(t)\,dt = \int_0^x u(t)dt. \tag{42}$$

Example 3. Find $\displaystyle\frac{d}{dx}\int_0^x u(t)\,dt.$

Proceeding as before, we find that

$$\frac{d}{dx}\int_0^x u(t)\,dt = u(x). \tag{43}$$

We now turn to our main goal to reduce Volterra integral equation to an equivalent differential equation. This can be easily achieved by differentiating both sides of the integral equation, noting that Leibnitz rule should be used in differentiating the integral as stated above. The differentiating process should be continued as many times as needed until we obtain a pure differential equation with the integral sign removed. Moreover, the initial conditions needed can be obtained by substituting $x = 0$ in the integral equation and the resulting integro-differential equations as will be shown.

We are now ready to give the the following illustrative examples.

Example 4. Find the initial value problem equivalent to the integral equation

$$u(x) = 1 + \int_0^x u(t)dt. \tag{44}$$

Differentiating both sides of the integral equation and using Leibnitz rule yield

$$u'(x) = u(x). \tag{45}$$

The initial condition can be obtained by substituting $x = 0$ in both sides of the integral equation; hence we find $u(0) = 1$. Consequently, the corresponding initial value problem of first order is given by

$$u'(x) - u(x) = 0, \quad u(0) = 1. \tag{46}$$

Example 5. Reduce the following integral equation to an initial value problem

$$u(x) = x + \int_0^x (t - x)u(t)dt. \tag{47}$$

Differentiating both sides of the integral equation we obtain

$$u'(x) = 1 - \int_0^x u(t)dt. \tag{48}$$

We differentiate both sides of the integro-differential equation (48) to remove the integral sign, therefore we obtain

$$u''(x) = -u(x), \tag{49}$$

or equivalently

$$u''(x) + u(x) = 0. \tag{50}$$

The related initial conditions are obtained by substituting $x = 0$ in $u(x)$ and $u'(x)$ in the equations above, and as a result we find $u(0) = 0$ and $u'(0) = 1$. Combining the above results yields the equivalent initial value problem of the second order given by

$$u''(x) + u(x) = 0, \ \ u(0) = 0, u'(0) = 1, \tag{51}$$

with constant coefficients that can be easily handled.

Example 6. Find the initial value problem equivalent to the integral equation

$$u(x) = x^3 + \int_0^x (x-t)^2 u(t)\, dt. \tag{52}$$

Differentiating both sides of (52) three times we find

$$\begin{cases} u'(x) &=& 3x^2 + 2\int_0^x (x-t)u(t)dt, \\ u''(x) &=& 6x + 2\int_0^x u(t)dt, \\ u'''(x) &=& 6 + 2u(x). \end{cases} \tag{53}$$

The proper initial conditions can be easily obtained by substituting $x = 0$ in $u(x)$, $u'(x)$ and $u''(x)$ in the equations above. Consequently, we obtain the nonhomogeneous initial value problem of third order given by

$$u'''(x) - 2u(x) = 6, \ \ u(0) = u'(0) = u''(0) = 0, \tag{54}$$

with constant coefficients that can be easily solved.

We point out here that the solution of initial value problems, that result from transforming Volterra integral equations, will be discussed in Chapter 3.

Exercises 1.4

In exercise 1 – 4, find $\frac{d}{dx}$ for the given integrals by using Leibnitz rule:

(1) $\int_0^x (x-t)^3 u(t)\,dt$

(2) $\int_x^{x^2} e^{xt}\,dt$

(3) $\int_0^x (x-t)^4 u(t)\,dt$

(4) $\int_x^{4x} \sin(x+t)\,dt$

In exercise 5 – 12, reduce each of the Volterra integral equations to an equivalent initial value problem:

(5) $u(x) = 1 + x + \int_0^x (x-t)^2 u(t)\,dt$

(6) $u(x) = e^x - \int_0^x (x-t)u(t)\,dt$

(7) $u(x) = x + \int_0^x (x-t)u(t)\,dt$

(8) $u(x) = x - \cos x + \int_0^x (x-t)u(t)\,dt$

(9) $u(x) = 2 + 3x + 5x^2 + \int_0^x [1 + 2(x-t)]u(t)\,dt$

(10) $u(x) = -5 + 6x + \int_0^x (5 - 6x + 6t)u(t)\,dt$

(11) $u(x) = \tan x - \int_0^x u(t)\,dt,\ x < \pi/2$

(12) $u(x) = 1 + x + \frac{5}{2}x^2 + \int_0^x [3 + 6(x-t) - \frac{5}{2}(x-t)^2]u(t)\,dt$

(13) $u(x) = x^4 + x^2 + 2\int_0^x (x-t)^2 u(t)\,dt$

(14) $u(x) = x^2 + \frac{1}{6}\int_0^x (x-t)^3 u(t)\,dt$

1.5 Converting IVP to Volterra Equation

In this section, we will study the method that converts an initial value problem to an equivalent Volterra integral equation. Before outlining the method needed, we wish to recall the useful transformation formula

$$\int_0^x \int_0^{x_1} \int_0^{x_2} \cdots \int_0^{x_{n-1}} f(x_n)\, dx_n \cdots dx_1 = \frac{1}{(n-1)!} \int_0^x (x-t)^{n-1} f(t)\, dt, \tag{55}$$

that converts any multiple integral to a single integral. This is an essential and useful formula that will be employed in the method that will be used in the conversion technique. We point out that this formula appears in most calculus texts. For practical considerations, the formulas

$$\int_0^x \int_0^x f(t) dt dt = \int_0^x (x-t) f(t)\, dt, \tag{56}$$

and

$$\int_0^x \int_0^x \int_0^x f(t) dt dt dt = \frac{1}{2!} \int_0^x (x-t)^2 f(t)\, dt, \tag{57}$$

are two special cases of the formula given above, and the mostly used formulas that will transform double and triple integrals respectively to a single integral for each. For simplicity reasons, we prove the first formula (56) that converts double integral to a single integral. Noting that the right hand side of (56) is a function of x allows us to set the equation

$$I(x) = \int_0^x (x-t) f(t)\, dt. \tag{58}$$

Differentiating both sides of (58), and using Leibnitz rule we obtain

$$I'(x) = \int_0^x f(t)\, dt. \tag{59}$$

Integrating both sides of (59) from 0 to x, noting that $I(0) = 0$ from (58), we find

$$I(x) = \int_0^x \int_0^x f(t)\, dt dt. \tag{60}$$

Equating the right hand sides of (58) and (60) completes the proof for this special case. The proof for the conversion of the triple integral to a single integral given by (57) may be carried out in the same manner. The procedure of reducing multiple integral to a single integral will be illustrated by examining the following examples.

Example 1. Convert the following quadruple integral

$$I(x) = \int_0^x \int_0^x \int_0^x \int_0^x u(t)\, dtdtdtdt, \tag{61}$$

to a single integral. Using the formula (55), noting that $n = 4$, we find

$$I(x) = \frac{1}{3!} \int_0^x (x-t)^3 u(t)\, dt, \tag{62}$$

the equivalent single integral.

Example 2. Convert the triple integral

$$I(x) = \int_0^x \int_0^x \int_0^x u(t)\, dtdtdt, \tag{63}$$

to a single integral. Using the formula (55)yields

$$I(x) = \frac{1}{2!} \int_0^x (x-t)^2 u(t)\, dt, \tag{64}$$

the equivalent single integral.

Returning to the main goal of this section, we discuss the technique that will be used to convert an initial value problem to an equivalent Volterra integral equation. Without loss of generality, and for simplicity reasons, we apply this technique to a third order initial value problem given by

$$y'''(x) + p(x)y''(x) + q(x)y'(x) + r(x)y(x) = g(x) \tag{65}$$

subject to the initial conditions

$$y(0) = \alpha, y'(0) = \beta, y''(0) = \gamma, \quad \alpha,\ \beta \text{ and } \gamma \text{ are constants.} \tag{66}$$

The coefficient functions $p(x)$, $q(x)$ and $r(x)$ are analytic functions by assuming that these functions have Taylor expansions about the origin. Besides, we assume that $g(x)$ is continuous through the interval of discussion. To transform (65) into an equivalent Volterra integral equation, we first set

$$y'''(x) = u(x), \tag{67}$$

where u(x) is a continuous function on the interval of discussion. Based on (67), it remains to find other relations for y and its derivatives as single integrals involving $u(x)$. This can be simply performed by integrating both sides of (67) from 0 to x where we find

$$y''(x) - y''(0) = \int_0^x u(t)dt, \tag{68}$$

or equivalently

$$y''(x) = \gamma + \int_0^x u(t)dt, \tag{69}$$

obtained upon using the initial condition $y''(0) = \gamma$. To obtain $y'(x)$ we integrate both sides of (69) from 0 to x to find that

$$y'(x) = \beta + \gamma x + \int_0^x \int_0^x u(t)dtdt. \tag{70}$$

Similarly we integrate both sides of (70) from 0 to x to obtain

$$y(x) = \alpha + \beta x + \frac{1}{2}\gamma x^2 + \int_0^x \int_0^x \int_0^x u(t)dtdtdt. \tag{71}$$

Using the conversion formulas (56) and (57), to reduce the double and triple integrals in (70) and (71) respectively to single integrals yields

$$y'(x) = \beta + \gamma x + \int_0^x (x-t)u(t)dt, \tag{72}$$

and

$$y(x) = \alpha + \beta x + \frac{1}{2}\gamma x^2 + \frac{1}{2}\int_0^x (x-t)^2 u(t)dt, \tag{73}$$

respectively. Substituting (67), (69), (72) and (73) into (65) leads to the following Volterra integral equation of the second kind

$$u(x) = f(x) + \int_0^x K(x,t)u(t)dt, \tag{74}$$

where

$$K(x,t) = p(x) + q(x)(x-t) + \frac{1}{2!}r(x)(x-t)^2, \tag{75}$$

and

$$f(x) = g(x) - \left\{\gamma p(x) + \beta q(x) + \alpha r(x) + \gamma x q(x) + r(x)\left(\beta x + \frac{1}{2}\gamma x^2\right)\right\}. \tag{76}$$

The following examples will be used to illustrate the above discussed technique.

Example 3. Convert the following initial value problem

$$y''' - 3y'' - 6y' + 5y = 0, \tag{77}$$

subject to the initial conditions

$$y(0) = y'(0) = y''(0) = 1, \tag{78}$$

to an equivalent Volterra integral equation.

As indicated before, we first set

$$y'''(x) = u(x). \tag{79}$$

Integrating both sides of (79) from 0 to x and using the initial condition $y''(0) = 1$ we find

$$y''(x) = 1 + \int_0^x u(t)dt. \tag{80}$$

Integrating (80) twice and using the proper initial conditions we find

$$y'(x) = 1 + x + \int_0^x \int_0^x u(t)dtdt \tag{81}$$

and

$$y(x) = 1 + x + \frac{1}{2}x^2 + \int_0^x \int_0^x \int_0^x u(t)dtdtdt. \tag{82}$$

Transforming the double and triple integrals in (81) and (82) to single integrals by using the formulas (56) and (57) we find

$$y'(x) = 1 + x + \int_0^x (x-t)u(t)dt, \tag{83}$$

and

$$y(x) = 1 + x + \frac{1}{2}x^2 + \frac{1}{2}\int_0^x (x-t)^2 u(t)dt. \qquad (84)$$

Substituting (79), (80), (83) and (84) into (77) we find

$$u(x) = 4 + x - \frac{5}{2}x^2 + \int_0^x \left(3 + 6(x-t) - \frac{5}{2}(x-t)^2\right) u(t)dt, \qquad (85)$$

the equivalent Volterra integral equation.

Example 4. Find the equivalent Volterra integral equation to the following initial value problem

$$y''(x) + y(x) = \cos x, \; y(0) = 0, \; y'(0) = 1. \qquad (86)$$

Proceeding as before, we set

$$y''(x) = u(x). \qquad (87)$$

Integrating both sides of (87) from 0 to x, using the initial condition $y'(0) = 1$ yields

$$y'(x) = 1 + \int_0^x u(t)dt. \qquad (88)$$

Integrating (88), using the initial condition $y(0) = 0$ leads to

$$y(x) = x + \int_0^x \int_0^x u(t)\, dtdt, \qquad (89)$$

or equivalently

$$y(x) = x + \int_0^x (x-t)u(t)\, dt, \qquad (90)$$

upon using the conversion rule (56). Inserting (87) and (90) into (86) leads to the following required Volterra integral equation

$$u(x) = \cos x - x - \int_0^x (x-t)u(t)\, dt, \qquad (91)$$

the equivalent Volterra integral equation.

As previously remarked, linear Volterra integral equations will be discussed extensively in Chapter 3. It is of interest to point out that the newly developed methods and the traditional methods will be introduced in that chapter.

Exercises 1.5

In exercises 1 – 3, convert each of the following first order initial value problem to a Volterra integral equation:

(1) $y' + y = 0, \; y(0) = 1$

(2) $y' - y = x, \; y(0) = 0$

(3) $y' + y = \sec^2 x, \; y(0) = 0$

In exercises 4 – 10, derive an equivalent Volterra integral equation to each of the following initial value problems of second order:

(4) $y'' + y = 0, \; y(0) = 1, y'(0) = 0$

(5) $y'' - y = 0, \; y(0) = 1, y'(0) = 1$

(6) $y'' + 5y' + 6y = 0, \; y(0) = 1, \; y'(0) = 1$

(7) $y'' + y' = 0, \; y(1) = 0, \; y'(1) = 1$

(8) $y'' + y' - 2y = 2x, \; y(0) = 0, \; y'(0) = 1$

(9) $y'' + y = \sin x, \; y(0) = 0, \; y'(0) = 0$

(10) $y'' - \sin x \, y' + e^x y = x, \; y(0) = 1, \; y'(0) = -1$

In exercises 11 – 15, convert each of the following initial value problems of higher order to an equivalent Volterra integral equation:

(11) $y''' - y'' - y' + y = 0, \; y(0) = 2, \; y'(0) = 0, \; y''(0) = 2$

(12) $y''' + 4y' = x, \; y(0) = 0, \; y'(0) = 0, \; y''(0) = 1$

(13) $y^{iv}+2y''+y=3x+4,\ y(0)=0, y'(0)=0, y''(0)=1, y'''(0)=1$

(14) $y^{iv}-y=0,\ y(0)=1,\ y'(0)=0,\ y''(0)=-1, y'''(0)=0$

(15) $y^{iv}+y''=2e^{x},\ y(0)=2,\ y'(0)=2,\ y''(0)=1, y'''(0)=1$

1.6 Converting BVP to Fredholm Equation

So far we have discussed how an initial value problem can be transformed to an equivalent Volterra integral equation. In this section, we will present the technique that will be used to convert a boundary value problem to an equivalent Fredholm integral equation. The technique is similar to that discussed in the previous section with some exceptions that are related to the boundary conditions. It is important to point out here that the procedure of reducing boundary value problem to Fredholm integral equation is complicated and rarely used. The method is similar to the technique discussed above, that reduces initial value problem to Volterra integral equation, with the exception that we are given boundary conditions.

A special attention should be taken to define $y'(0)$, since it is not always given, as will be seen later. This can be easily determined from the resulting equations. It seems useful and practical to illustrate this method by applying it to an example rather than proving it.

Example 1. We want to derive an equivalent Fredholm integral equation to the following boundary value problem

$$y''(x)+y(x)=x,\ 0<x<\pi, \tag{92}$$

subject to the boundary conditions

$$y(0)=1,\ y(\pi)=\pi-1. \tag{93}$$

We first set

$$y''(x)=u(x). \tag{94}$$

Integrating both sides of (94) from 0 to x gives

$$\int_0^x y''(t)dt=\int_0^x u(t)\,dt, \tag{95}$$

or equivalently

$$y'(x) = y'(0) + \int_0^x u(t)\,dt. \tag{96}$$

As indicated earlier, $y'(0)$ is not given in this boundary value problem. However, $y'(0)$ will be determined later by using the boundary condition at $x = \pi$.

Integrating both sides of (96) from 0 to x and using the given boundary condition at $x = 0$ we find

$$y(x) = 1 + xy'(0) + \int_0^x (x-t)u(t)\,dt, \tag{97}$$

upon converting the resulting double integral to a single integral as discussed before. It remains to evaluate $y'(0)$, and this can be obtained by substituting $x = \pi$ in both sides of (97) and using the boundary condition at $x = \pi$, hence we find

$$y(\pi) = 1 + \pi y'(0) + \int_0^\pi (\pi - t)u(t)\,dt. \tag{98}$$

Solving (98) for $y'(0)$ we obtain

$$y'(0) = \frac{1}{\pi}\left((\pi - 2) - \int_0^\pi (\pi - t)\,u(t)\,dt\right). \tag{99}$$

Substituting (99) for $y'(0)$ into (97) yields

$$y(x) = 1 + \frac{x}{\pi}\left((\pi - 2) - \int_0^\pi (\pi - t)u(t)dt\right) + \int_0^x (x-t)u(t)\,dt. \tag{100}$$

Substituting (94) and (100) into (92) we get

$$\begin{aligned} u(x) &= x - 1 - \frac{x}{\pi}\left((\pi - 2) - \int_0^\pi (\pi - t)u(t)dt\right) \\ &\quad - \int_0^x (x-t)u(t)\,dt. \end{aligned} \tag{101}$$

The following identity

$$\int_0^\pi (.) = \int_0^x (.) + \int_x^\pi (.), \tag{102}$$

will carry the equation (101) to

$$\begin{aligned} u(x) &= x - 1 - \frac{x}{\pi}(\pi - 2) + \frac{x}{\pi}\int_0^x (\pi - t)u(t)dt \\ &\quad + \frac{x}{\pi}\int_x^\pi (\pi - t)u(t)\,dt - \int_0^x (x - t)u(t)dt, \end{aligned} \tag{103}$$

or equivalently, after performing simple calculations and adding integrals with similar limits

$$u(x) = \frac{2x - \pi}{\pi} - \int_0^x \frac{t(x - \pi)}{\pi}u(t)dt - \int_x^\pi \frac{x(t - \pi)}{\pi}u(t)dt. \tag{104}$$

Consequently, the desired Fredholm integral equation of the second kind is given by

$$u(x) = \frac{2x - \pi}{\pi} - \int_0^\pi K(x, t)u(t)\,dt, \tag{105}$$

where the kernel $K(x, t)$ is defined by

$$K(x, t) = \begin{cases} \dfrac{t(x - \pi)}{\pi} & \text{for } 0 \leq t \leq x \\ \dfrac{x(t - \pi)}{\pi} & \text{for } x \leq t \leq \pi \end{cases} \tag{106}$$

It is worth noting that the equation (105) obtained is a nonhomogeneous Fredholm integral equation, and this usually results when converting a nonhomogeneous boundary value problem to its equivalent integral equation. However, homogeneous boundary value problems always lead to homogeneous Fredholm integral equations. Further, we point out here that the solution of boundary value problems is much easier if compared with the solution of its corresponding Fredholm integral equation. This leads to the conclusion that transforming boundary value problem to Fredholm integral equation is less important if compared with transforming initial value problems to Volterra integral equations.

It is of interest to note that the recent developed methods, namely the Adomian decomposition method and the direct computation method, will be introduced in Chapter 2 to handle Fredholm integral Equations. In addition, the traditional methods, namely the successive approximations method and the method of successive substitutions, will be used in Chapter 2 as well .

Exercises 1.6

Derive the equivalent Fredholm integral equation for each of the following boundary value problems:

1. $y'' + 4y = \sin x,\ 0 < x < 1, \quad y(0) = y(1) = 0$

2. $y'' + 2xy = 1,\ 0 < x < 1, \quad y(0) = y(1) = 0$

3. $y'' + y = x,\ 0 < x < 1, \quad y(0) = 1, y(1) = 0$

4. $y'' + y = x,\ 0 < x < 1, \quad y(0) = 1, y'(1) = 0$

Chapter 2

Fredholm Integral Equations

2.1 Introduction

In this chapter we shall be concerned with the nonhomogeneous Fredholm integral equations of the second kind of the form

$$u(x) = f(x) + \lambda \int_a^b K(x,t)u(t)dt, \quad a \leq x \leq b, \tag{1}$$

where $K(x,t)$ is the kernel of the integral equation, and λ is a parameter. A considerable amount of discussion will be directed towards the various methods and techniques that are used for solving this type of equations starting with the most recent methods that proved to be highly reliable and accurate. To do this we will naturally focus our study on the *degenerate* or *separable* kernels all through this chapter. The standard form of the *degenerate* or *separable* kernel is given by

$$K(x,t) = \sum_{k=1}^{n} g_k(x)\, h_k(t). \tag{2}$$

The expressions $x - t$, $x + t, xt$, $x^2 - 3xt + t^2$, etc. are examples of separable kernels. For other well-behaved non-separable kernels, we can convert it to separable in the form (2) simply by expanding these kernels using Taylor's expansion.

Moreover, the kernel $K(x,t)$ is defined to be square integrable in both x and t in the square $a \leq x \leq b$, $a \leq t \leq b$ if the following *regularity condition*

$$\int_a^b \int_a^b K(x,t)\, dx\, dt < \infty, \tag{3}$$

is satisfied. This condition gives rise to the development of the solution of the Fredholm integral equation (1).

It is also convenient to state,without proof, the so called *Fredholm Alternative Theorem* that relates the solutions of homogeneous and nonhomogeneous Fredholm integral equations: The nonhomogeneous Fredholm integral equation (1) has one and only one solution if the only solution to the homogeneous Fredholm integral equation

$$u(x) = \lambda \int_a^b K(x,t)u(t)dt, \tag{4}$$

is the trivial solution $u(x) = 0$. For more details about the *regularity condition* and the *Fredholm Alternative Theorem* the reader is referred to [7], [12], [13] and [14].

We end this section by introducing the necessary condition [16] that will guarantee a unique solution to the integral equation (1) in the interval of discussion. Considering (2), if the kernel $K(x,t)$ is real, continuous and bounded in the square $a \leq x \leq b$ and $a \leq t \leq b$, i.e. if

$$|K(x,t)| \leq M, \quad a \leq x \leq b, \quad \text{and} \quad a \leq t \leq b, \tag{5}$$

and if $f(x) \neq 0$, and continuous in $a \leq x \leq b$, then the necessary condition that will guarantee that (1) has only a unique solution is given by

$$|\lambda|\, M\, (b-a) < 1. \tag{6}$$

It is important to note that a continuous solution to Fredholm integral equation may exist [16], even though the condition (6) is not satisfied. This may be clearly seen by considering the equation

$$u(x) = -4 + \int_0^1 (2x + 3t)u(t)dt. \tag{7}$$

In this example, $\lambda = 1, |K(x,t)| \leq 5$ and $(b-a) = 1$; therefore

$$|\lambda| M (b-a) = 5 \not< 1. \tag{8}$$

Accordingly, the necessary condition (6) fails to hold, but in fact the integral equation (7) has an exact solution given by

$$u(x) = 4x, \tag{9}$$

and this can be justified through direct substitution.

As indicated in our objective for a first course in integral equations, we will pay more attention to the practical techniques for solving integral equations rather than the abstract theorems. In the following we will discuss several methods that handle successfully the Fredholm integral equations of the second kind starting with the most recent methods as indicated earlier.

2.2 The Decomposition Method

Adomian [1] recently developed the so-called Adomian decomposition method or simply the *decomposition method*. The method was well introduced by Adomian in his recent books [1] and [2] and several related bibliography [3] and [4] for example. The method proved to be reliable and effective for a wide class of equations, differential and integral equations, linear and nonlinear models. The method provides the solution in a series form as will be seen soon. The method was applied mostly to ordinary and partial differential equations, and was rarely used for integral equations in [1] and [2]. The concept of convergence of the solution obtained by this method was addressed by [1] and [2] and extensively by [8] and [9] for nonlinear problems. The convergence concept is beyond the scope of this text. However, the decomposition method can be successfully applied towards linear and nonlinear integral equations.

In the decomposition method we usually express the solution $u(x)$ of the integral equation (1) in a series form defined by

$$u(x) = \sum_{n=0}^{\infty} u_n(x). \tag{10}$$

Substituting the decomposition (10) into both sides of (1) yields

$$\sum_{n=0}^{\infty} u_n(x) = f(x) + \lambda \int_a^b K(x,t) \left(\sum_{n=0}^{\infty} u_n(t) \right) dt, \tag{11}$$

or equivalently

$$\begin{aligned} u_0(x) + u_1(x) + u_2(x) + u_3(x) + \cdots = f(x) &+ \lambda \int_a^b K(x,t) u_0(t) dt \\ &+ \lambda \int_a^b K(x,t)\, u_1(t) dt \\ &+ \lambda \int_a^b K(x,t)\, u_2(t) dt \\ &+ \cdots \end{aligned} \tag{12}$$

The components $u_0(x), u_1(x), u_2(x), u_3(x), \ldots$ of the unknown function $u(x)$ are completely determined in a recurrent manner if we set

$$u_0(x) = f(x), \tag{13}$$

$$u_1(x) = \lambda \int_a^b K(x,t) u_0(t) dt, \tag{14}$$

$$u_2(x) = \lambda \int_a^b K(x,t) u_1(t) dt, \tag{15}$$

$$u_3(x) = \lambda \int_a^b K(x,t) u_2(t) dt, \tag{16}$$

and so on. The above discussed scheme for the determination of the components $u_0(x), u_1(x), u_2(x), u_3(x), \ldots$ of the solution $u(x)$ of Eq. (1) can be written in a recursive manner by

$$u_0(x) = f(x), \tag{17}$$

$$u_{n+1}(x) = \lambda \int_a^b K(x,t) u_n(t) dt, \; n \geq 0. \tag{18}$$

In view of (17) and (18), the components $u_0(x)$, $u_1(x)$, $u_2(x)$, $u_3(x)$, $u_4(x)$...follow immediately. With these components determined, the solution $u(x)$ of (1) is readily determined in a series form using the decomposition (10). It is important to note that the series obtained for $u(x)$ frequently provides the exact solution in a closed form as will be illustrated later. However, for concrete problems, where (10) cannot be evaluated, a truncated series $\sum_{n=0}^{k} u_n(x)$ is usually used to approximate the solution $u(x)$ if a numerical solution is desired. We point out here that few terms of the truncated series usually provide the higher accuracy level of the approximate solution if compared with the existing numerical techniques. The decomposition technique proved to be effective and reliable even if applied to nonlinear Fredholm integral equations as will be discussed in Chapter 6.

In the following we discuss some examples that illustrate the decomposition method outlined above.

Example 1. We first consider the Fredholm integral equation of the second kind

$$u(x) = \frac{9}{10}x^2 + \int_0^1 \frac{1}{2}\, x^2\, t^2\, u(t)\, dt. \tag{19}$$

It is clear that $f(x) = \frac{9}{10}x^2$, $\lambda = 1, K(x,t) = \frac{1}{2}x^2t^2$. To evaluate the components $u_0(x)$, $u_1(x)$, $u_2(x), \ldots$ of the series solution, we use the recursive scheme (17) and (18) to find

$$\begin{aligned}
u_0(x) &= \frac{9}{10}x^2, && (20)\\
u_1(x) &= \int_0^1 \frac{1}{2}\, x^2\, t^2 u_0(t) dt, \\
&= \int_0^1 \frac{1}{2}\, x^2\, \frac{9}{10} t^4 dt, \\
&= \frac{9}{100}x^2, && (21)\\
u_2(x) &= \int_0^1 \frac{1}{2}x^2t^2u_1(t)dt, \\
&= \int_0^1 \frac{1}{2}x^2t^2\frac{9}{100}t^2, dt, \\
&= \frac{9}{1000}x^2, && (22)
\end{aligned}$$

and so on. Noting that

$$u(x) = u_0(x) + u_1(x) + u_2(x) + \cdots, \tag{23}$$

we can easily obtain the solution in a series form given by

$$u(x) = \frac{9}{10}x^2 + \frac{9}{100}x^2 + \frac{9}{1000}x^2 + \cdots, \tag{24}$$

so that the solution of (19) in a closed form

$$u(x) = x^2, \tag{25}$$

follows immediately upon using the formula for the sum of the infinite geometric series.

Example 2. We next consider the Fredholm integral equation

$$u(x) = \cos x + 2x + \int_0^{\pi} xtu(t)\, dt. \tag{26}$$

Proceeding as in Example 1, we set

$$u_0(x) = \cos x + 2x, \tag{27}$$

$$\begin{aligned} u_1(x) &= \int_0^{\pi} xtu_0(t)dt, \\ &= \int_0^{\pi} xt\,(\cos t + 2t)\, dt, \\ &= \left(-2 + \frac{2}{3}\pi^3\right) x \end{aligned} \tag{28}$$

$$\begin{aligned} u_2(x) &= \int_0^{\pi} xtu_1(t)dt, \\ &= \int_0^{\pi} x\left(-2 + \frac{2}{3}\pi^3\right) t^2 dt, \\ &= \left(-\frac{2}{3}\pi^3 + \frac{2}{9}\pi^6\right) x. \end{aligned} \tag{29}$$

Consequently, the solution of (26) in a series form is given by

$$u(x) = \cos x + 2x + \left(-2 + \frac{2}{3}\pi^3\right) x + \left(-\frac{2}{3}\pi^3 + \frac{2}{9}\pi^6\right) x + \cdots \tag{30}$$

and in a closed form

$$u(x) = \cos x, \tag{31}$$

by eliminating the so-called self-cancelling noise terms between various components of $u(x)$. The answer obtained can be justified through substitution. The self-cancelling noise terms are defined to be similar terms with opposite signs that will vanish in the limit.

Example 3. We consider here the Fredholm integral equation

$$u(x) = e^x - 1 + \int_0^1 t\,u(t)\,dt. \tag{32}$$

Applying the decomposition technique as discussed before we find

$$\begin{aligned}
u_0(x) &= e^x - 1, && (33)\\
u_1(x) &= \int_0^1 t u_0(t) dt, \\
&= \int_0^1 t\left(e^t - 1\right) dt, \\
&= \frac{1}{2}, && (34)\\
u_2(x) &= \int_0^1 t u_1(t) dt, \\
&= \int_0^1 \frac{1}{2} t\, dt, \\
&= \frac{1}{4}. && (35)
\end{aligned}$$

The determination of the components (33) – (35) yields the solution of the equation (32) in a series form given by

$$u(x) = e^x - 1 + \frac{1}{2}\left(1 + \frac{1}{2} + \frac{1}{4} + \cdots\right), \tag{36}$$

where we can easily obtain the solution in a closed form given by

$$u(x) = e^x, \tag{37}$$

by evaluating the sum of the infinite geometric series in the right hand side of Eq. (36).

It is important to note that the evaluation of the components $u_0(x)$, $u_1(x), u_2(x), \ldots$ is simple as we observed from the examples above. However, we can still reduce the size of calculations by using a modified version of the decomposition method. In this modified approach, we often need to evaluate the first two components $u_0(x)$ and $u_1(x)$ only. In the following, we explain how we can employ the easy modified algorithm.

2.2.1 The Modified Decomposition Method:

It is worthnoting that the decomposition method may be sometimes implemented in a different but easier manner. It is recommended to apply the modified decomposition method for cases where the nonhomogeneous part $f(x)$ in (1) consists of a polynomial that includes many terms, or in the case $f(x)$ contains a combination of polynomial and other trigonometric or transcendental functions. This modified technique, as will be seen later, will minimize the volume of calculations and reduce the several integral evaluations that result in applying the standard decomposition method.

It is also of interest, before giving a clear discussion of this method, to note that this modified technique will be carried out with promising results in Volterra integral equations and nonlinear integral equations in Chapters 3 and 6. The technique avoids the cumbersome integrations of other methods.

It is clear in this modified method , we simply split the given function $f(x)$ into two parts defined by

$$f(x) = f_1(x) + f_2(x), \tag{38}$$

where $f_1(x)$ consists of one term of $f(x)$ in many problems or two terms for other cases, and $f_2(x)$ includes the remaining terms of $f(x)$. We note that a necessary condition is required to apply this approach in that $f(x)$ should consist of more than one term as shown by (38). In view of (38), the integral equation (1) becomes

$$u(x) = f_1(x) + f_2(x) + \lambda \int_a^b K(x,t)u(t)dt, \quad a \leq x \leq b. \tag{39}$$

Substituting the decomposition given by (10) into (39) and using few terms of the expansion we obtain

$$\begin{aligned}u_0(x) + u_1(x) + u_2(x) + \cdots = f_1(x) + f_2(x) & +\lambda \int_a^b K(x,t)u_0(t)dt \\ & +\lambda \int_a^b K(x,t)\, u_1(t)dt \\ & +\lambda \int_a^b K(x,t)\, u_2(t)dt \\ & + \cdots \end{aligned} \tag{40}$$

The components $u_0(x), u_1(x), u_2(x), u_3(x), ...$ of the unknown function $u(x)$ can be completely determined in a recurrent manner if we assign $f_1(x)$ only to the zeroth component $u_0(x)$, whereas the function $f_2(x)$ will be added to the formula of the component $u_1(x)$ given before in Eq. (14). In other words the modified decomposition method works elegantly if we set

$$u_0(x) = f_1(x), \tag{41}$$

$$u_1(x) = f_2(x) + \lambda \int_a^b K(x,t)u_0(t)dt, \tag{42}$$

$$u_2(x) = \lambda \int_a^b K(x,t)u_1(t)dt, \tag{43}$$

$$u_3(x) = \lambda \int_a^b K(x,t)u_2(t)dt, \tag{44}$$

and so on. The above discussed scheme for the determination of the components $u_0(x), u_1(x), u_2(x), u_3(x), ...$ of the solution $u(x)$ of the equation (1) can be written in a recursive manner by

$$u_0(x) = f_1(x), \tag{45}$$

$$u_1(x) = f_2(x) + \lambda \int_a^b K(x,t)u_0(t)dt, \tag{46}$$

$$u_{n+1}(x) = \lambda \int_a^b K(x,t)u_n(t)dt, \; n \geq 1. \tag{47}$$

Recall that in most problems we need to use (45) and (46) only.

The modified decomposition scheme can be explained by the following illustrative examples:

Example 4. We consider here the Fredholm integral equation

$$u(x) = e^{3x} - \frac{1}{9}\left(2e^3 + 1\right)x + \int_0^1 xt\,u(t)\,dt. \tag{48}$$

To apply the modified decomposition scheme as dicussed above, we first split the function $f(x)$ into

$$f_1(x) = e^{3x}, \tag{49}$$

and

$$f_2(x) = -\frac{1}{9}\left(2e^3 + 1\right)x. \tag{50}$$

Therefore, we set

$$u_0(x) = e^{3x}, \tag{51}$$

and

$$\begin{aligned} u_1(x) &= -\tfrac{1}{9}\left(2e^3+1\right)x + \int_0^1 xtu_0(t)dt, \\ &= -\tfrac{1}{9}\left(2e^3+1\right)x + x\int_0^1 te^{3t}dt, \\ &= 0. \end{aligned} \tag{52}$$

In view of (52) , we conclude that $u_n = 0$, $n \geq 1$. The exact solution

$$u(x) = e^{3x}, \tag{53}$$

follows immediately.

Example 5. We consider here the Fredholm integral equation

$$u(x) = \sin^{-1}x + \left(\frac{\pi}{2} - 1\right)x - \int_0^1 x\,u(t)\,dt. \tag{54}$$

Applying the modified decomposition method as dicussed above, we first split the function $f(x)$ into

$$f_1(x) = \sin^{-1}x, \tag{55}$$

and

$$f_2(x) = \left(\frac{\pi}{2} - 1\right) x. \tag{56}$$

Therefore, we set

$$\begin{aligned} u_0(x) &= \sin^{-1} x, &(57)\\ u_1(x) &= \left(\frac{\pi}{2} - 1\right) x - \int_0^1 x u_0(t)dt, \\ &= \left(\frac{\pi}{2} - 1\right) x - x\int_0^1 \sin^{-1} t\, dt, \\ &= 0. &(58) \end{aligned}$$

Consequently, the components $u_n(x) = 0,\ n \geq 1$. The exact solution

$$u(x) = \sin^{-1} x, \tag{59}$$

is readily obtained.

This confirms our belief that the decomposition method and the modified decomposition method introduce the solution of Fredholm integral equation in the form of a rapidly convergent power series with elegantly computable terms.

Exercises 2.2

In exercises 1 – 12, solve the following Fredholm integral equations by using the *decomposition method*

1. $u(x) = \frac{13}{3}x - \frac{1}{4}\int_0^1 xtu(t)dt.$

2. $u(x) = x^3 - \frac{1}{5}x + \int_0^1 xtu(t)dt.$

3. $u(x) = x^2 + \int_0^1 xtu(t)dt.$

4. $u(x) = e^x + e^{-1}\int_0^1 u(t)dt.$

5. $u(x) = x + \sin x - x\int_0^{\pi/2} u(t)dt.$

6. $u(x) = x + \cos x - 2x\int_0^{\pi/6} u(t)dt.$

7. $u(x) = \cos(4x) + \frac{1}{4}x - \int_0^{\pi/8} xu(t)dt.$

8. $u(x) = \sinh x - e^{-1}x + \int_0^1 xtu(t)dt.$

9. $u(x) = 2e^{2x} + \left(1 - e^2\right)x + \int_0^1 xu(t)dt.$

10. $u(x) = 1 + \sec^2 x - \int_0^{\pi/4} u(t)dt.$

11. $u(x) = \sin x + \int_{-1}^1 e^{\sin^{-1} x}u(t)dt.$

12. $u(x) = \tan x - \int_{-\pi/3}^{\pi/3} e^{\tan^{-1} x}u(t)dt.$

In exercises 13 – 20 solve the given Fredholm integral equations by using the *modified decomposition method*:

13. $u(x) = \tan^{-1} x + \frac{1}{2}\left(\ln 2 - \frac{\pi}{2}\right)x + \int_0^1 xu(t)dt.$

14. $u(x) = \cosh x + (\sinh 1)x + \left(e^{-1} - 1\right) - \int_0^1 (x-t)u(t)dt.$

15. $u(x) = \frac{1}{1+x^2} + 2x\sinh(\pi/4) - x\int_{-1}^1 e^{\tan^{-1} t}u(t)dt.$

16. $u(x) = \frac{1}{\sqrt{1-x^2}} + \left(e^{\pi/6} - 1\right)x - x\int_0^{1/2} e^{\sin^{-1} t}u(t)dt.$

17. $u(x) = \frac{1}{1+x^2} + \frac{\pi^2}{32}x - x\int_0^1 \tan^{-1} t u(t)dt.$

18. $u(x) = \cos^{-1} x - \pi x + \int_{-1}^1 xu(t)dt.$

19. $u(x) = x\tan^{-1} x + \left(\frac{\pi}{4} - \frac{1}{2}\right)x - \int_0^1 xu(t)dt.$

20. $u(x) = x\sin^{-1} x + 1 - \left(\frac{\pi}{8} + 1\right)x + \int_0^1 xu(t)dt.$

Hint: $f_1(x) = x\sin^{-1} x + 1.$

2.3 The Direct Computation Method

We next introduce an efficient method for solving Fredholm integral equations of the second kind (1). Recall that our attention will be focused on separable or degenerate kernels K(x,t) expressed in the form defined by (2). Without loss of generality and for simplicity reasons, we may assume that the kernel of (1) can be expressed as

$$K(x,t) = g(x)h(t). \tag{60}$$

Accordingly, the equation (1) becomes

$$u(x) = f(x) + \lambda g(x) \int_a^b h(t)u(t)dt. \tag{61}$$

It is clear that the definite integral at the right hand side of (61) reveals that the integrand depends on one variable, namely the variable t. This means that the definite integral in the right hand side of (61) is equivalent to a numerical value α, where α is a constant. In other words, we may write

$$\alpha = \int_a^b h(t)u(t)dt. \tag{62}$$

It follows that equation (61) becomes

$$u(x) = f(x) + \lambda \alpha g(x). \tag{63}$$

It is thus obvious that the solution $u(x)$ is completely determined by (63) upon evaluating the constant α. This can be easily done by substituting Eq. (63) into Eq. (62). We point out here that this approach [24] is slightly different than other existing techniques in that we substitute (63) into (62) and not in (61) as used by other texts.

It is worthnoting that the *direct computation method* determines the exact solution in a closed form, rather than a series form, provided that the constant α is evaluated. In addition, this method usually gives rise to a system of algebraic equations depending on the structure of the kernel, where sometimes we need to evaluate more than one constant as will be seen in Example 2. In Chapter 6, the computational difficulty may arise in determining the constant α if the resulting algebraic equation is of third degree or higher. This kind of difficulty may arise in nonlinear integral equations.

The above discussed technique will be explained by the following illustrative examples.

Example 1. We will use the direct computation method to solve the following Fredholm integral equation

$$u(x) = \frac{5}{6}x + \frac{1}{2}\int_0^1 xt\,u(t)dt. \tag{64}$$

As indicated before we set

$$\alpha = \int_0^1 t\,u(t)dt, \tag{65}$$

where α is a constant that represents the numerical value of the integral (65). The equation (65) carries (64) into

$$u(x) = \left(\frac{5}{6} + \frac{1}{2}\alpha\right)x. \tag{66}$$

To determine α, we substitute (66) into (65) to obtain

$$\alpha = \int_0^1 \left(\frac{5}{6} + \frac{1}{2}\alpha\right)t^2 dt, \tag{67}$$

so that by integrating the right hand side and solving for α we find

$$\alpha = \frac{1}{3}. \tag{68}$$

Substituting (68) into (66) yields

$$u(x) = x, \tag{69}$$

the exact solution of the given integral equation.

Example 2. We next solve the Fredholm integral equation

$$u(x) = -8x - 6x^2 + \int_0^1 \left(20xt^2 + 12x^2t\right)u(t)dt, \tag{70}$$

by using the direct computation method. Noting that the kernel here is separable and consists of two terms, we can rewrite Eq. (70) as

$$u(x) = -8x - 6x^2 + 20x\int_0^1 t^2u(t)dt + 12x^2\int_0^1 tu(t)dt. \tag{71}$$

In a manner parallel to the preceding example, we set

$$\alpha = \int_0^1 t^2u(t)dt, \tag{72}$$

and

$$\beta = \int_0^1 tu(t)dt, \tag{73}$$

where α and β are constants. Consequently, Eq. (71) can be expressed in the form

$$u(x) = (20\alpha - 8)\,x + (12\beta - 6)\,x^2. \tag{74}$$

Substituting (74) into (72) and (73) we obtain

$$\alpha = \int_0^1 \left[(20\alpha - 8)\,t + (12\beta - 6)\,t^2\right] t^2dt, \tag{75}$$

and

$$\beta = \int_0^1 \left[(20\alpha - 8)\,t + (12\beta - 6)\,t^2\right] tdt. \tag{76}$$

Integrating the right hand side of equations (75) and (76) yields the system of equations

$$5\alpha + 3\beta = 4, \tag{77}$$

$$40\alpha + 12\beta = 25, \tag{78}$$

so that by solving this system we find

$$\alpha = \frac{9}{20},\quad \beta = \frac{7}{12}. \tag{79}$$

Inserting (79) into (74) gives

$$u(x) = x^2 + x, \tag{80}$$

the exact solution of the integral equation under discussion.

Example 3. Now we solve the Fredholm integral equation

$$u(x) = \sec^2 x - 1 + \int_0^{\frac{\pi}{4}} u(t)dt, \tag{81}$$

by using the direct computation method. Proceeding as before we set

$$\alpha = \int_0^{\frac{\pi}{4}} u(t)dt, \tag{82}$$

and by using this into (81) yields

$$u(x) = \sec^2 x - 1 + \alpha. \tag{83}$$

Inserting (83) into (82) we find

$$\alpha = \int_0^{\frac{\pi}{4}} \left(\sec^2 t - 1 + \alpha\right) dt, \tag{84}$$

so that

$$\alpha = 1. \tag{85}$$

Substituting (85) into (83) gives

$$u(x) = \sec^2 x, \tag{86}$$

the exact solution of the integral equation of Example 3.

In closing this section, we point out that the direct computation method introduces a very direct technique to formally determine the solution of Fredholm integral equation. In this method, the Fredholm integral equation will be transformed into a more readily solvable integral. Moreover, the direct computation method introduces the exact solution in a closed form rather than a series form as in the case of the decomposition method. The other traditional methods, that will be discussed in the forthcoming sections, also determine the solution in a series form, but in a different approach than the decomposition method. We remark here that the direct computation method was introduced in this section in a slightly different manner than other texts.

Exercises 2.3

Solve the following Fredholm integral equations by using the *direct computation method*:

1. $u(x) = xe^x - x + \int_0^1 xu(t)dt.$

2. $u(x) = x^2 - \frac{25}{12}x + 1 + \int_0^1 xtu(t)dt.$

3. $u(x) = x\sin x - x + \int_0^{\pi/2} xu(t)dt.$

4. $u(x) = e^{2x} - \frac{1}{4}\left(e^2 + 1\right)x + \int_0^1 xtu(t)dt.$

5. $u(x) = \sec^2 x - \frac{\pi}{4} + \int_0^{\pi/4} u(t)dt.$

6. $u(x) = \sin(2x) - \frac{1}{2}x + \int_0^{\pi/4} xu(t)dt.$

7. $u(x) = x^2 - \frac{1}{3}x - \frac{1}{4} + \int_0^1 (x+2)\,u(t)dt.$

8. $u(x) = \sin x + \cos x - \frac{\pi}{2}x + \int_0^{\pi/2} xtu(t)dt.$

9. $u(x) = \sec x\tan x + x - \int_0^{\pi/3} xu(t)dt.$

10. $u(x) = x^2 - \frac{1}{6}x - \frac{1}{24} + \frac{1}{2}\int_0^1 (1 + x - t)\,u(t)dt.$

11. $u(x) = \sin x - \frac{x}{4} + \frac{1}{4}\int_0^{\pi/2} xtu(t)dt.$

12. $u(x) = 1 + \int_{0+}^1 \ln(xt)u(t)dt, \quad 0 < x \leq 1.$

13. $u(x) = \frac{9}{10}x^3 + \frac{1}{2}\int_0^1 x^3tu(t)dt.$

14. $u(x) = 1 + \frac{1}{2}\int_0^{\pi/4} \sec^2 x\, u(t)dt.$

2.4 The Successive Approximations Method

In this method, we replace the unknown function under the integral sign of the Fredholm integral equation of the second kind

$$u(x) = f(x) + \lambda \int_a^b K(x,t)u(t)dt, \quad a \leq x \leq b, \tag{87}$$

by any *selective* real valued function $u_0(x)$, $a \leq x \leq b$. Accordingly, the first approximation $u_1(x)$ of the solution $u(x)$ is defined by

$$u_1(x) = f(x) + \lambda \int_a^b K(x,t)u_0(t)dt. \tag{88}$$

The second approximation of $u_2(x)$ of the solution $u(x)$ can be obtained by replacing $u_0(x)$ in (88) by the obtained approximation $u_1(x)$, hence we find

$$u_2(x) = f(x) + \lambda \int_a^b K(x,t)u_1(t)dt. \tag{89}$$

This process can be continued in the same manner to obtain the *nth* approximation. In other words, the various approximations of the solution $u(x)$ of (87) can be obtained in a recursive scheme given by

$$\begin{cases} u_0(x) &= \text{any selective real valued function} \\ u_n(x) &= f(x) + \lambda \int_a^b K(x,t)u_{n-1}(t)dt, \quad n \geq 1. \end{cases} \tag{90}$$

Even though we can select any real valued function for the zeroth approximation $u_0(x)$, the most commonly selected functions for $u_0(x)$ are 0,1 or x. At the limit, the solution $u(x)$ is obtained by

$$u(x) = \lim_{n \to \infty} u_n(x), \tag{91}$$

so that the resulting solution $u(x)$ is independent of the choice of $u_0(x)$.

It is important to distinguish between the recursive schemes used in the decomposition method and in the successive approximations method. In the decomposition method, we apply the approach to determine several components of the solution $u(x)$ where, in this case

$$u(x) = \sum_{n=0}^{\infty} u_n(x), \tag{92}$$

so that the zeroth component $u_0(x)$ is defined by all terms that are out of the integral sign or part of these terms if the modified version is used. However, in the successive approximations method, we apply the above recursive scheme (90) to determine various approximations of the solution $u(x)$ itself, and not components of $u(x)$. Further, we should note here that the zeroth approximation $u_0(x)$ is not defined but rather given by a selective function, and as a result the solution $u(x)$ is given by the formula (91).

The successive approximations method will be illustrated by the following examples.

Example 1. Consider the Fredholm integral equation

$$u(x) = e^x + e^{-1} \int_0^1 u(t)\, dt. \tag{93}$$

As indicated above we can select any real value function for the zeroth component, hence we set

$$u_0(x) = 0. \tag{94}$$

Substituting (94) into the right hand side of (93) we find

$$u_1(x) = e^x + e^{-1} \int_0^1 u_0(t)\, dt, \tag{95}$$

and this gives the first approximation of $u(x)$ by

$$u_1(x) = e^x. \tag{96}$$

Inserting (96) into (95) to replace $u_0(x)$ we obtain

$$u_2(x) = e^x + e^{-1} \int_0^1 e^t\, dt, \tag{97}$$

where by integration we determine the second approximation of $u(x)$ by

$$u_2(x) = e^x + 1 - e^{-1}. \tag{98}$$

Continuing in the same manner we find the third approximation of $u(x)$ given by

$$u_3(x) = e^x + 1 - e^{-2}. \tag{99}$$

Proceeding as before, we obtain the nth component

$$u_n(x) = e^x + 1 - e^{-(n-1)}, \quad n \geq 1. \tag{100}$$

Using (91) , the solution $u(x)$ of (93) is given by

$$\begin{aligned} u(x) &= \lim_{n\to\infty} u_n(x), \\ &= \lim_{n\to\infty} \left(e^x + 1 - e^{-(n-1)}\right) \\ &= e^x + 1, \end{aligned} \tag{101}$$

obtained upon evaluating the limit as $n \to \infty$.

Example 2. We next consider the Fredholm integral equation

$$u(x) = x + \lambda \int_0^1 xtu(t)dt. \tag{102}$$

The zeroth approximation may by selected by

$$u_0(x) = 0, \tag{103}$$

where by substituting this in the right hand side of (102) the first approximation

$$u_1(x) = x, \tag{104}$$

follows immediately. Proceeding in the same manner we find that

$$u_2(x) = x + \lambda \int_0^1 xt^2 dt, \tag{105}$$

so that

$$u_2(x) = x + \frac{\lambda}{3}x. \tag{106}$$

In a similar manner we obtain

$$u_3(x) = x + \lambda \int_0^1 xt\left(1 + \frac{\lambda}{3}\right) tdt, \tag{107}$$

which yields

$$u_3(x) = x + \frac{\lambda}{3}x + \frac{\lambda^2}{9}x. \tag{108}$$

Generally we obtain for the nth approximation

$$u_n(x) = x + \frac{\lambda}{3}x + \frac{\lambda^2}{9}x + \cdots + \frac{\lambda^{n-1}}{3^{n-1}}x, \quad n \geq 1. \tag{109}$$

Consequently, the solution $u(x)$ of (102) is given by

$$\begin{aligned} u(x) &= \lim_{n\to\infty} u_n(x), \\ &= \lim_{n\to\infty}\left(x + \tfrac{\lambda}{3}x + \tfrac{\lambda^2}{9}x + \cdots\right) \\ &= \tfrac{3}{3-\lambda}x, \quad 0 < \lambda < 3. \end{aligned} \tag{110}$$

To show that $u(x)$ obtained in (110) does not depend on the selection of $u_0(x)$, we will solve the equation (102) by selecting

$$u_0(x) = x. \tag{111}$$

Using the new selection of $u_0(x)$ in the right hand side of (102) the first approximation

$$u_1(x) = x + \frac{\lambda}{3}x, \tag{112}$$

is readily obtained. Proceeding as before we thus obtain

$$u_2(x) = x + \lambda \int_0^1 xt\left(t + \frac{\lambda}{3}t\right) dt, \tag{113}$$

which gives

$$u_2(x) = x + \frac{\lambda}{3}x + \frac{\lambda^2}{9}x. \tag{114}$$

In a parallel manner we find

$$u_n(x) = x + \frac{\lambda}{3}x + \frac{\lambda^2}{3^2}x + \cdots + \frac{\lambda^n}{3^n}x, \quad n \geq 1. \tag{115}$$

Accordingly, we obtain

$$u(x) = \frac{3}{3-\lambda}x, \quad 0 < \lambda < 3, \tag{116}$$

which is consistent with the same result obtained above in (110).

Exercises 2.4

Solve the following Fredholm integral equations by using the *successive approximations method*:

1. $u(x) = \frac{11}{12}x + \frac{1}{4}\int_0^1 xtu(t)dt.$
2. $u(x) = \frac{6}{7}x^3 + \frac{5}{7}\int_0^1 x^3tu(t)dt.$
3. $u(x) = \frac{13}{3}x - \frac{1}{4}\int_0^1 xtu(t)dt.$
4. $u(x) = 1 + \int_0^1 xu(t)dt.$
5. $u(x) = \sin x + \int_0^{\pi/2} \sin x \cos t\, u(t)dt.$
6. $u(x) = -\frac{1}{2} + \sec^2 x + \frac{1}{2}\int_0^{\pi/4} u(t)dt.$
7. $u(x) = -\frac{1}{4} + \sec x \tan x + \frac{1}{4}\int_0^{\pi/3} u(t)dt.$
8. $u(x) = \cosh x + \left(1 - e^{-1}\right)x - \int_0^1 xtu(t)dt.$
9. $u(x) = e^x - (\sinh 1)\, x + \frac{1}{2}\int_{-1}^1 xu(t)dt.$
10. $u(x) = \frac{1}{4}x + \sin x - \frac{1}{4}\int_0^{\pi/2} xu(t)dt.$

2.5 The Method of Successive Substitutions

This method introduces the solution of the integral equation in a series form through evaluating single integral and multiple integrals as well. The computational work needed in this method is huge compared with other techniques.

In this method, we set $x = t$ and $t = t_1$ in the Fredholm integral equation

$$u(x) = f(x) + \lambda \int_a^b K(x,t)u(t)dt, \quad a \le x \le b, \tag{117}$$

to obtain

$$u(t) = f(t) + \lambda \int_a^b K(t,t_1)u(t_1)dt_1. \tag{118}$$

Replacing $u(t)$ in the right hand side of (117) by its obtained value given by (118) yields

$$\begin{aligned} u(x) \quad &= f(x) \quad + \lambda \int_a^b K(x,t)f(t)dt \\ &\quad + \lambda^2 \int_a^b K(x,t) \int_a^b K(t,t_1)u(t_1)dt_1dt. \end{aligned} \tag{119}$$

Substituting $x = t_1$ and $t = t_2$ in (117) we obtain

$$u(t_1) = f(t_1) + \lambda \int_a^b K(t_1,t_2)u(t_2)dt_2. \tag{120}$$

Substituting the value of $u(t_1)$ obtained in (120) into the right hand side of (119) leads to

$$\begin{aligned} u(x) \quad &= f(x) \quad + \lambda \int_a^b K(x,t)f(t)dt \\ &\quad + \lambda^2 \int_a^b \int_a^b K(x,t)K(t,t_1)f(t_1)dt_1dt \\ &\quad + \lambda^3 \int_a^b \int_a^b \int_a^b K(x,t)K(t,t_1)K(t_1,t_2)u(t_2)dt_2dt_1dt. \end{aligned} \tag{121}$$

Accordingly, the general series form for $u(x)$ can be written as

$$\begin{aligned} u(x) \quad &= f(x) \quad + \lambda \int_a^b K(x,t)f(t)dt \\ &\quad + \lambda^2 \int_a^b \int_a^b K(x,t)K(t,t_1)f(t_1)dt_1dt \\ &\quad + \lambda^3 \int_a^b \int_a^b \int_a^b K(x,t)K(t,t_1)K(t_1,t_2)f(t_2)dt_2dt_1dt \\ &\quad + \cdots \end{aligned} \tag{122}$$

We note that the series solution given in (122) converges uniformly in the interval $[a, b]$ if $\lambda M(b-a) < 1$ where $|K(x,t)| \leq M$. The proof of the theorem appears in the texts [16], [19], [20] and others. We remark here that in this method the unknown function $u(x)$ is replaced by the given function $f(x)$ that makes the evaluation of the several multiple integrals possible and easily computable. This substitution of $u(x)$ occurs several times through the integrals and hence this is why it is called the method of successive substitutions. The technique will be illustrated by discussing the following examples.

Example 1. We solve the following Fredholm integral equation

$$u(x) = \frac{23}{6}x + \frac{1}{8}\int_0^1 xtu(t)dt, \tag{123}$$

by using the method of successive substitutions.
Substituting $\lambda = \frac{1}{8}$, $f(x) = \frac{23}{6}x$, and $K(x,t) = xt$ into (122) yields

$$u(x) = \frac{23}{6}x + \frac{1}{8}\int_0^1 \frac{23}{6}xt^2dt + \frac{1}{8^2}\int_0^1\int_0^1 \frac{23}{6}x{t_1}^2t^2dt_1dt + \cdots, \tag{124}$$

or equivalently

$$u(x) = \frac{23}{6}x\left[1 + \frac{1}{24} + \frac{1}{576} + \cdots\right], \tag{125}$$

so that we obtain the solution

$$u(x) = 4x, \tag{126}$$

upon evaluating the sum of the geometric series.

Example 2. We next solve the Fredholm integral equation

$$u(x) = 1 + \frac{1}{4}\int_0^{\pi/2} \cos x\, u(t)dt, \tag{127}$$

by using the method of successive substitutions.
Substituting $\lambda = \frac{1}{4}$, $f(x) = 1$, and $K(x,t) = \cos x$ into (122) yields

$$u(x) = 1 + \frac{1}{4}\int_0^{\pi/2} \cos x\, dt + \frac{1}{16}\int_0^{\pi/2}\int_0^{\pi/2} \cos x\, \cos t\, dt_1\, dt + \cdots, \tag{128}$$

and this will yield

$$u(x) = 1 + \left(\frac{\pi}{8}\cos x + \frac{\pi}{32}\cos x + \cdots\right), \tag{129}$$

which gives the exact solution

$$u(x) = 1 + \frac{\pi}{6}\cos x, \tag{130}$$

obtained upon using the sum of the infinite geometric series.

Exercises 2.5

Solve the following Fredholm integral equations by using the *successive substitutions method*:

1. $u(x) = \frac{11}{6}x + \frac{1}{4}\int_0^1 xtu(t)dt.$
2. $u(x) = 1 - \frac{1}{4}\int_0^{\pi/2} \cos xu(t)dt.$
3. $u(x) = \frac{7}{12}x + 1 + \frac{1}{2}\int_0^1 xtu(t)dt.$
4. $u(x) = \cos x + \frac{1}{2}\int_0^{\pi/2} \sin xu(t)dt.$
5. $u(x) = \frac{7}{8}x^2 + \frac{1}{2}\int_0^1 x^2tu(t)dt.$
6. $u(x) = \frac{9}{10}x^3 + \frac{1}{2}\int_0^1 x^3tu(t)dt.$
7. $u(x) = \sin x + \frac{1}{2}\int_0^{\pi/2} \cos xu(t)dt.$
8. $u(x) = 1 + \frac{1}{2}\int_0^{\pi/2} \sin xu(t)dt.$
9. $u(x) = 1 + \frac{1}{2}\int_0^{\pi/4} \sec^2 x\, u(t)dt.$
10. $u(x) = 1 + \frac{1}{5}\int_0^{\pi/3} \sec x \tan x\, u(t)dt.$

2.6 Comparison between Alternative Methods

Having finished the mathematical analysis of the methods that handle Fredholm integral equations, we are now ready to carry out a comparison between these methods. When it comes to select a preferable method among the four methods for solving linear Fredholm integral equations, we cannot recommend a specific method. However, we found that if the separable kernel $K(x,t)$ of the integral equation consists of a polynomial of one or two terms only, the *direct computation method* might be the best choice because it provides the exact solution with the minimum volume of calculations. For other types of kernels, and if in addition the nonhomogeneous part $f(x)$ is a polynomial of more than two terms we found that the *decomposition method* proved to be effective, reliable and produces a rapid convergent series for the solution. The series obtained by using the decomposition method may give the solution in a closed form or we may obtain an approximation of high accuracy level by using a truncated series for concrete problems.

It is worth noting that the decomposition method expands the solution $u(x)$ about a function, instead of a point as in Taylor theorem.

To compare the *decomposition method* with the *successive approximation method*, it is clear that the decomposition method is easier in that we integrate always few terms to obtain the successive components, whereas in the other method we integrate many terms to evaluate the successive approximations after selecting the zeroth approximation. The two methods give the solution in a series form.

In addition, we point out that the *method of successive substitutions* suffers from the huge size of calculations in evaluating the several multiple integrals especially if the function $f(x)$ is a trigonometric, logarithmic or exponential function. However, the method is directly based on substituting the unknown function $u(x)$ under the integral sign by the given function $f(x)$.

It is to be noted that, for a first course in integral equations, we introduced four methods only to handle Fredholm integral equations, noting that other traditional techniques are left for a further study.

To achieve our goal of the comparison between these methods, we demonstrate this comparison by discussing the following example by using all various methods.

Example 1. We solve the following example

$$u(x) = \frac{5}{6}x + \frac{1}{2}\int_0^1 xtu(t)dt, \tag{131}$$

by using the four alternative methods discussed before.

(a) The Decomposition Method: In this method, we have to set the zeroth component u_0 by all terms outside the integral sign, hence we have

$$u_0(x) = \frac{5}{6}x. \tag{132}$$

Using (132) we obtain the first component $u_1(x)$ by

$$u_1(x) = \frac{1}{2}x\int_0^1 \frac{5}{6}t^2 dt, \tag{133}$$

so that

$$u_1(x) = \frac{5}{6^2}x. \tag{134}$$

Proceeding in the same manner we can easily obtain

$$u_2(x) = \frac{5}{6^3}x. \tag{135}$$

Noting that in the decomposition method we have

$$u(x) = u_0 + u_1 + u_2 + u_3 + \cdots, \tag{136}$$

so that

$$u(x) = \frac{5}{6}x\left(1 + \frac{1}{6} + \frac{1}{6^2} + \frac{1}{6^3} + \cdots\right), \tag{137}$$

and this gives the exact solution

$$u(x) = x, \tag{138}$$

obtained by evaluating the sum of the infinite geometric series.

(b) The Direct Computation Method: As discussed earlier we set

$$\alpha = \int_0^1 t\,u(t)dt, \tag{139}$$

where α is a constant that represents the numerical value of the integral (139). The equation (139) carries (131) into

$$u(x) = \left(\frac{5}{6} + \frac{1}{2}\alpha\right) x. \tag{140}$$

To determine α, we substitute (140) into (139) to obtain

$$\alpha = \int_0^1 \left(\frac{5}{6} + \frac{1}{2}\alpha\right) t^2 dt, \tag{141}$$

so that by integrating the right hand side and solving for α we find

$$\alpha = \frac{1}{3}. \tag{142}$$

Substituting (142) into (140) yields

$$u(x) = x, \tag{143}$$

the exact solution of the equation (131).

(c) The Successive Approximations Method: In this method we select the zeroth approximation by

$$u_0(x) = 0. \tag{144}$$

Following the technique that was discussed above, the other approximations of the solution $u(x)$ can be easily obtained by

$$u_1(x) = \frac{5}{6}x, \tag{145}$$

$$u_2(x) = \left(\frac{5}{6} + \frac{5}{6^2}\right) x, \tag{146}$$

$$u_3(x) = \left(\frac{5}{6} + \frac{5}{6^2} + \frac{5}{6^3}\right) x. \tag{147}$$

and so on. Accordingly, the nth component is given by

$$u_n(x) = \left(\frac{5}{6} + \frac{5}{6^2} + \frac{5}{6^3} + \frac{5}{6^4} + \cdots + \frac{5}{6^n}\right) x, \quad n \geq 1. \tag{148}$$

Consequently we find

$$\begin{aligned} u(x) &= \lim_{n\to\infty} u_n(x), \\ &= \lim_{n\to\infty} \tfrac{5}{6}x\left(1+\tfrac{1}{6}+\tfrac{1}{6^2}+\tfrac{1}{6^3}+\tfrac{1}{6^4}+\cdots\right) \\ &= x, \end{aligned} \tag{149}$$

the same result obtained above.

(d) The Method of Successive Substitutions: In this method we have to set $K(x,t)=xt$, $\lambda=\frac{1}{2}$ and $f(x)=\frac{5}{6}x$, hence we have

$$\begin{aligned} u(x) &= \frac{5}{6}x+\frac{1}{2}\int_0^1 \frac{5}{6}xt^2dt+\frac{1}{4}\int_0^1\int_0^1 \frac{5}{6}x^2tt_1{}^2dt_1dt+\cdots \\ &= \frac{5}{6}x\left(1+\frac{1}{6}+\frac{1}{6^2}+\cdots\right) \\ &= x. \end{aligned} \tag{150}$$

This confirms our belief that the Adomian decomposition method and the direct computation method reduce the size of calculations and provide improvements if compared with other traditional techniques.

2.7 Homogeneous Fredholm Equations

In this section we will study the homogeneous Fredholm integral equation with separable kernel given by

$$u(x)=\lambda\int_a^b K(x,t)u(t)\,dt, \tag{151}$$

obtained from (1) by setting $f(x)=0$. It is easily seen that the trivial solution $u(x)=0$ is a solution of the homogeneous Fredholm integral equation (151). In this study our goal will be focused on finding nontrivial solutions to (151) if exist. We can achieve our goal by introducing the technique that will enable us to determine the nontrivial solutions to (151). Generally speaking, the homogeneous Fredholm integral equation with separable kernel may have nontrivial solutions. Our approach in obtaining these desired solutions will be based mainly on the *direct computation method* that was employed effectively for nonhomogeneous Fredholm integral equations. We point out that Adomian decomposition method is not applicable for the homogeneous Fredholm integral equations. This may be related to the fact that the nonhomogeneous part $f(x)$ does not exist in this type of problems, and therefore the zeroth component $u_0(x)$ cannot be defined.

We recall that the direct computation method reduces the equation to an algebraic equation if the kernel consists of one term only, or to a system of algebraic equations if the kernel contains many separable terms. Additional discussions will be required for determining possible values of λ that will give rise to the nontrivial solutions as will be discussed soon.

Without loss of generality, we may assume a one term kernel given by

$$K(x,t) = g(x)\,h(t), \tag{152}$$

so that (151) becomes

$$u(x) = \lambda g(x) \int_a^b h(t)u(t)\,dt. \tag{153}$$

Using the direct computation method we set

$$\alpha = \int_a^b h(t)u(t)\,dt, \tag{154}$$

so that (153) becomes

$$u(x) = \lambda \alpha g(x). \tag{155}$$

We note that $\alpha = 0$ gives the trivial solution $u(x) = 0$, by using Eq. (155), which is not our desired goal in this study. However, to determine the nontrivial solutions of (151), we need to determine the values of the parameter λ by considering $\alpha \neq 0$. This can be done by substituting (155) into (154) to obtain

$$\alpha = \lambda \alpha \int_a^b h(t)g(t)dt, \tag{156}$$

or equivalently

$$1 = \lambda \int_a^b h(t)g(t)dt \tag{157}$$

which gives a numerical value for $\lambda \neq 0$ by evaluating the definite integral in (157). Having evaluated λ, the nontrivial solution given by (155) is determined.

For separable kernels that contain more than one term, the method reduces the homogeneous Fredholm integral equation to a system of algebraic equations as will be seen from the examples below.

In closing this section we point out that the particular nonzero values of λ that result from solving the algebraic system of equations are called the *eigenvalues* of the kernel. Moreover, substituting the obtained values of λ in (155) gives the usually called *eigenfunctions* of the equation which are the nontrivial solutions of (151).

The following illustrative examples will be used to explain the technique introduced above and the concept of *eigenvalues* and *eigenfunctions.*

Example 1. We first solve the homogeneous Fredholm integral equation with one term kernel

$$u(x) = \lambda \int_0^1 x\, u(t)dt. \tag{158}$$

As indicated above , (158) becomes

$$u(x) = \lambda \alpha x, \tag{159}$$

where

$$\alpha = \int_0^1 u(t)dt. \tag{160}$$

Substituting (159) into (160) yields

$$\alpha = \lambda\alpha \int_0^1 tdt, \tag{161}$$

which gives

$$\alpha = \frac{1}{2}\lambda\alpha, \tag{162}$$

so that the eigenvalue of the kernel

$$\lambda = 2, \tag{163}$$

obtained by noting that $\alpha \neq 0$. Substituting (163) into (162) yields

$$\alpha = \alpha, \tag{164}$$

which indicates that α is an arbitrary constant . Using (163) and (164) in (159) leads to the eigenfunction of the equation given by

$$u(x) = 2\alpha x, \tag{165}$$

obtained upon using (159).

Example 2. We next solve the homogeneous Fredholm integral equation

$$u(x) = \frac{2}{\pi}\lambda \int_0^{\pi} \cos(x + t)u(t)dt. \tag{166}$$

The equation (166) can be rewritten as

$$u(x) = \frac{2}{\pi}\lambda \cos x \int_0^{\pi} \cos tu(t)dt - \frac{2}{\pi}\lambda \sin x \int_0^{\pi} \sin tu(t)dt, \tag{167}$$

or by

$$u(x) = \frac{2}{\pi}\lambda \left(\alpha \cos x - \beta \sin x\right), \tag{168}$$

where

$$\alpha = \int_0^{\pi} \cos t\, u(t)\, dt, \tag{169}$$

$$\beta = \int_0^{\pi} \sin t\, u(t)\, dt, \tag{170}$$

Substituting (168) into (169) and (170) and integrating yield

$$\alpha = \lambda\alpha, \tag{171}$$

and

$$\beta = -\lambda\beta. \tag{172}$$

For $\alpha \neq 0$ and $\beta \neq 0$, we obtain the eigenvalues

$$\lambda_1 = 1, \quad \text{and} \quad \lambda_2 = -1. \tag{173}$$

Substituting $\lambda_1 = 1$ in (171) and (172) yields

$$\alpha = \alpha, \quad \text{and} \quad \beta = 0, \tag{174}$$

which gives the eigenfunction corresponding to $\lambda_1 = 1$ by

$$u_1(x) = \frac{2}{\pi}\alpha \cos x, \tag{175}$$

obtained upon using (168), where α is an arbitrary constant.
Similarly, substituting $\lambda_2 = -1$ in (171) and (172) yields

$$\alpha = 0, \quad \text{and} \quad \beta = \beta, \tag{176}$$

which gives the second eigenfunction corresponding to $\lambda_2 = -1$ by

$$u_2(x) = \frac{2}{\pi}\beta \sin x, \tag{177}$$

obtained upon using (168), where β is an arbitrary constant.

Example 3. We finally solve the homogeneous Fredholm integral equation with two terms kernel

$$u(x) = \lambda \int_0^1 (6x - 2t)u(t)dt. \tag{178}$$

Equation (178) can be rewritten as

$$u(x) = 6\lambda\alpha x - \beta\lambda, \tag{179}$$

where

$$\alpha = \int_0^1 u(t)\, dt, \tag{180}$$

$$\beta = \int_0^1 2tu(t)\, dt, \tag{181}$$

Substituting (179) into (180) and (181) and integrating yield

$$(1-3\lambda)\alpha+\lambda\beta=0, \tag{182}$$

and

$$-4\lambda\alpha+(1+\lambda)\beta=0. \tag{183}$$

For $\alpha \neq 0$ and $\beta \neq 0$, we obtain the eigenvalues

$$\lambda_1=\lambda_2=1. \tag{184}$$

Substituting $\lambda_1 = 1$ in (182) and (183) yields

$$\beta=2\alpha, \tag{185}$$

Consequently the eigenfunctions corresponding to $\lambda_1 = \lambda_2 = 1$ are given by

$$u_1(x)=u_2(x)=6\alpha x-2\alpha, \tag{186}$$

obtained upon using (179) .

Exercises 2.7

Find the nontrivial solutions for following homogeneous Fredholm integral equations by using the *eigenvalues* and *eigenfunctions* concepts

1. $u(x)=\lambda\int_0^1 2tu(t)dt.$
2. $u(x)=\lambda\int_0^1 4xu(t)dt.$
3. $u(x)=\lambda\int_0^1 xe^t u(t)dt.$
4. $u(x)=\lambda\int_0^{\pi/2} \cos x\sin t\, u(t)dt.$
5. $u(x)=\frac{2}{\pi}\lambda\int_0^{\pi} \sin(x+t)u(t)dt.$
6. $u(x)=\frac{2}{\pi}\lambda\int_0^{\pi} \cos(x-t)u(t)dt.$

7. $u(x) = \lambda \int_0^{\pi/3} \sec x \tan t \, u(t) dt.$

8. $u(x) = \lambda \int_0^{\pi/4} \sec^2 x \, u(t) dt.$

9. $u(x) = \lambda \int_0^{1} \sin^{-1} x u(t) dt.$

10. $u(x) = \lambda \int_0^{1} (3 - \frac{3}{2}x) t u(t) dt.$

Chapter 3

Volterra Integral Equations

3.1 Introduction

In this chapter we shall be concerned with the nonhomogeneous Volterra integral equation of the second kind of the form

$$u(x) = f(x) + \lambda \int_0^x K(x,t)u(t)dt, \tag{1}$$

where $K(x,t)$ is the kernel of the integral equation, and λ is a parameter. As indicated earlier the limits of integration for the Volterra integral equations [22] are functions of x and not constants as in Fredholm integral equations. The kernel in equation (1) will be considered a separable kernel as discussed before in the previous chapter. Our concern will be on applying various methods to determine the solution $u(x)$ of (1) and not on the abstract theorems related to the existence, uniqueness of the solution or the convergence concept.

We discussed in Section 1.5 the technique that converts initial value problems to Volterra integral equations. In the following we will discuss several methods that handle successfully the linear Volterra integral equations in a manner parallel to our approach in discussing Chapter 2. Accordingly we will start with the recent methods.

3.2 The Adomian Decomposition Method

As mentioned earlier, Adomian recently developed the *Adomian decomposition method* or simply the *decomposition method* that proved to work for all types of differential, integral and integro-differential equations, linear or nonlinear. The method was introduced by Adomian in his books [1] and [2] and other related research papers such as [3] and [4]. The focus of the two books was mainly on ordinary and partial differential equations. We have seen from Chapter 2 that the decomposition method establishes the solution in the form of a power series. The approach we will follow here is identical to the same approach that was implemented earlier in Chapter 2. In this method $u(x)$ will be decomposed into components, that will be determined, given by the series form

$$u(x) = \sum_{n=0}^{\infty} u_n(x), \tag{2}$$

with u_0 identified as all terms out of the integral sign, i.e.

$$u_0(x) = f(x). \tag{3}$$

Substituting (2) into (1) yields

$$\sum_{n=0}^{\infty} u_n(x) = f(x) + \lambda \int_0^x K(x,t) \left(\sum_{n=0}^{\infty} u_n(t) \right) dt, \tag{4}$$

which by using few terms of the expansion gives

$$\begin{aligned} u_0(x) + u_1(x) + u_2(x) + \cdots = f(x) & + \lambda \int_0^x K(x,t) u_0(t) dt \\ & + \lambda \int_0^x K(x,t) u_1(t) dt \\ & + \lambda \int_0^x K(x,t) u_2(t) dt \\ & + \lambda \int_0^x K(x,t) u_3(t) dt \\ & + \cdots \end{aligned} \tag{5}$$

The components $u_0(x)$, $u_1(x)$, $u_2(x)$, $u_3(x)$, ... of the unknown function $u(x)$ are completely determined in a recurrent manner if we set

$$u_0(x) = f(x), \tag{6}$$

$$u_1(x) = \lambda \int_0^x K(x,t)u_0(t)dt, \tag{7}$$

$$u_2(x) = \lambda \int_0^x K(x,t)u_1(t)dt, \tag{8}$$

$$u_3(x) = \lambda \int_0^x K(x,t)u_2(t)dt, \tag{9}$$

and so on. The above discussed scheme for the determination of the components $u_0(x), u_1(x), u_2(x), u_3(x), \ldots$ of the solution $u(x)$ of Eq. (1) can be written in a recursive scheme by

$$u_0(x) = f(x), \tag{10}$$

$$u_{n+1}(x) = \lambda \int_0^x K(x,t)u_n(t)dt, \quad n \geq 0. \tag{11}$$

In view of (10) and (11), the components $u_0(x)$, $u_1(x)$, $u_2(x)$, $u_3(x)$, $u_4(x)$,...follow immediately upon integrating the easily computable integrals. With these components determined, the solution $u(x)$ of (1) is readily determined in a series form upon using (2). As discussed before, the series obtained for $u(x)$ frequently provides the exact solution in a closed form as will be illustrated later. However, for concrete problems, where (2) cannot be evaluated, a truncated series $\sum_{n=0}^{k} u_n(x)$ is usually used to approximate the solution $u(x)$. It is to be noted here that, for numerical purposes, few terms of the obtained series usually provide the higher accuracy level of the approximation of the solution if compared with the existing numerical techniques.

It is interesting to recall that the decomposition method provides the solution of any style of equations in the form of a power series with easily computable components. In addition, applications have shown a very fast convergence of the series solution. The convergence concept of the decomposition technique was addressed extensively by [8] and [9] and by others, but it will not be discussed in this text.

We indicated earlier, that the decomposition technique proved to be powerful and reliable even if applied to nonlinear Volterra integral equations as will be discussed in Chapter 6.

The following illustrative examples will be discussed to explain the above outlined decomposition method.

Example 1. We first consider the Volterra integral equation

$$u(x) = 1 + \int_0^x u(t)\,dt. \tag{12}$$

It is clear that $f(x) = 1$, $\lambda = 1, K(x,t) = 1$. Using the decomposition series solution (2) and the recursive scheme (10) and (11) to determine the components u_n, $n \geq 0$, we find

$$\begin{aligned} u_0(x) &= 1, &(13)\\ u_1(x) &= \int_0^x u_0(t)dt,\\ &= \int_0^x dt,\\ &= x, &(14)\\ u_2(x) &= \int_0^x u_1(t)dt,\\ &= \int_0^x tdt,\\ &= \frac{1}{2!}x^2, &(15) \end{aligned}$$

and so on. Noting that

$$u(x) = u_0(x) + u_1(x) + u_2(x) + \cdots, \tag{16}$$

we can easily obtain the solution in a series form given by

$$u(x) = 1 + x + \frac{1}{2!}x^2 + \cdots, \tag{17}$$

so that the solution in a closed form is

$$u(x) = e^x, \tag{18}$$

upon using the Taylor expansion for e^x .

Example 2. We next consider the Volterra integral equation

$$u(x) = x + \int_0^x (t - x)u(t)\,dt. \tag{19}$$

Proceeding as in Example 1, we set

$$u_0(x) = x, \tag{20}$$

$$\begin{aligned} u_1(x) &= \int_0^x (t-x)u_0(t)dt, \\ &= \int_0^x t\,(t-x)\,dt, \\ &= -\frac{1}{3!}x^3, \end{aligned} \tag{21}$$

$$\begin{aligned} u_2(x) &= \int_0^x (t-x)u_1(t)dt, \\ &= \int_0^x -\frac{1}{3!}t^3\,(t-x)\,dt, \\ &= \frac{1}{5!}x^5. \end{aligned} \tag{22}$$

Consequently, the solution of (19) in a series form is given by

$$u(x) = x - \frac{1}{3!}x^3 + \frac{1}{5!}x^5 + \cdots \tag{23}$$

and in a closed form by

$$u(x) = \sin x, \tag{24}$$

by using the Taylor expansion of $\sin x$.

Example 3. We consider here the Volterra integral equation

$$u(x) = 6x - x^3 + \frac{1}{2}\int_0^x t\,u(t)\,dt. \tag{25}$$

Applying the decomposition technique as discussed before we find

$$\begin{aligned} u_0(x) &= 6x - x^3, & (26)\\ u_1(x) &= \frac{1}{2}\int_0^x t u_0(t) dt, \\ &= \frac{1}{2}\int_0^x t\left(6t - t^3\right) dt, \\ &= x^3 - \frac{1}{10}x^5, & (27)\\ u_2(x) &= \frac{1}{2}\int_0^x t u_1(t) dt, \\ &= \frac{1}{2}\int_0^x t\left(t^3 - \frac{1}{10}t^5\right) dt, \\ &= \frac{1}{10}x^5 - \frac{1}{140}x^7. & (28) \end{aligned}$$

Consequently, the solution of (25) in a series form is given by

$$u(x) = \left(6x - x^3\right) + \left(x^3 - \frac{1}{10}x^5\right) + \left(\frac{1}{10}x^5 - \frac{1}{140}x^7\right) + \cdots, \quad (29)$$

where we can easily obtain the solution in a closed form given by

$$u(x) = 6x, \quad (30)$$

by eliminating the self-cancelling terms between various components of the solution $u(x)$.

It was discussed in Section 1.3 that it is not always possible to determine a solution in a closed form, but instead the solution obtained may be expressed in a series form. The difference between the closed form solution and the power series solution has been illustrated before in Section 1.3. However the series solution usually employed for numerical approximations, and the more terms we obtain provide more accuracy in the approximation of the solution. In the next example we discuss again the solution expressed in a series form.

Example 4. Consider the Volterra integral equation

$$u(x) = 1 + \frac{1}{2}\int_0^x x t^2 u(t)\, dt. \quad (31)$$

Following the procedure used above, we find

$$u_0(x) = 1, \tag{32}$$

$$\begin{aligned} u_1(x) &= \frac{1}{2}x\int_0^x t^2 dt, \\ &= \frac{1}{6}x^4, \end{aligned} \tag{33}$$

$$\begin{aligned} u_2(x) &= \frac{1}{12}x\int_0^x t^6 dt, \\ &= \frac{1}{84}x^8, \end{aligned} \tag{34}$$

$$\begin{aligned} u_3(x) &= \frac{1}{168}x\int_0^x t^{10} dt, \\ &= \frac{1}{1848}x^{12}. \end{aligned} \tag{35}$$

Consequently, the solution of (31) in a series form is given by

$$u(x) = 1 + \frac{1}{6}x^4 + \frac{1}{84}x^8 + \frac{1}{1848}x^{12} + \cdots, \tag{36}$$

so that a closed form for $u(x)$ does not appear obtainable.

Even though the decomposition method proved to be powerful and reliable, but it can be used sometimes in a more effective manner which we called the *modified decomposition method.* In a manner parallel to that used in Chapter 2, the volume of calculations will be reduced by evaluating only the first two components $u_0(x)$ and $u_1(x)$. The modified technique works for specific problems where the function $f(x)$ in (1) consists at least of two terms.

3.2.1 The Modified Decomposition Method:

It is important to note that the modified decomposition method, that was introduced before in Chapter 2 for the Fredholm integral equations, is also applicable here . In Volterra integral equations where the nonhomogeneous part $f(x)$ in (1) consists of a polynomial that includes many terms, or in the case $f(x)$ contains a combination of polynomial and other trigonometric or transcendental functions, the modified decomposition method showed to work extremely well. As indicated earlier, the technique may minimize the volume of calculations needed when applying the standard decomposition method. To achieve our goal, we decompose the function $f(x)$ into two parts such as

$$f(x) = f_1(x) + f_2(x), \tag{37}$$

where $f_1(x)$ consists of one term only, or if needed of two terms in fewer other cases, and $f_2(x)$ includes the remaining terms of $f(x)$. Accordingly, Eq. (1) becomes

$$u(x) = f_1(x) + f_2(x) + \lambda \int_0^x K(x,t)u(t)dt. \tag{38}$$

Substituting the decomposition given by (2) into (38) and using few terms of the expansions we obtain

$$\begin{aligned} u_0(x) + u_1(x) + u_2(x) + \cdots = f_1(x) + f_2(x) & + \lambda \int_0^x K(x,t)u_0(t)dt \\ & + \lambda \int_0^x K(x,t)\, u_1(t)dt \\ & + \lambda \int_0^x K(x,t)\, u_2(t)dt \\ & + \cdots \end{aligned} \tag{39}$$

Consequently, the components $u_0(x)$, $u_1(x)$, $u_2(x)$, $u_3(x), \ldots$ of the unknown function $u(x)$ can be completely determined in a recurrent manner if we assign $f_1(x)$ only to the component $u_0(x)$, whereas the component $f_2(x)$ will be added to the formula of the component $u_1(x)$ given before in Eq. (7). In other words, we can set

$$u_0(x) = f_1(x), \tag{40}$$

$$u_1(x) = f_2(x) + \lambda \int_0^x K(x,t)u_0(t)dt, \tag{41}$$

$$u_2(x) = \lambda \int_0^x K(x,t)u_1(t)dt, \tag{42}$$

$$u_3(x) = \lambda \int_0^x K(x,t)u_2(t)dt, \tag{43}$$

and so on. The method discussed above for the determination of the components $u_0(x), u_1(x), u_2(x), u_3(x), \cdots$ of the solution $u(x)$ of Eq. (1) can be written in a recursive relationship by

$$u_0(x) = f_1(x), \tag{44}$$

$$u_1(x) = f_2(x) + \lambda \int_0^x K(x,t)u_0(t)dt, \tag{45}$$

$$u_{n+1}(x) = \lambda \int_0^x K(x,t)u_n(t)dt, \; n \geq 1. \tag{46}$$

In most problems, we need to use (44) and (45) only.

For illustration purposes, we study the following example:

Example 5. We consider here the Volterra integral equation

$$u(x) = \sec x \tan x - \frac{1}{4}\left(e^{\sec x} - e\right)x + \frac{1}{4}\int_0^x x e^{\sec t} u(t)\,dt, \; x < \frac{\pi}{2}. \tag{47}$$

Using the modified decomposition method as dicussed above, we first decompose the function $f(x)$ into

$$f_1(x) = \sec x \tan x, \tag{48}$$

and

$$f_2(x) = -\frac{1}{4}\left(e^{\sec x} - e\right)x. \tag{49}$$

Consequently, we find

$$u_0(x) = \sec x \tan x, \tag{50}$$

$$\begin{aligned} u_1(x) &= -\frac{1}{4}\left(e^{\sec x} - e\right)x + \frac{1}{4}\int_0^x x e^{\sec t} u_0(t)\,dt, \\ &= -\frac{1}{4}\left(e^{\sec x} - e\right)x + \frac{1}{4}x\int_0^x \sec t \, \tan t \, e^{\sec t}\,dt, \\ &= 0, \end{aligned} \tag{51}$$

obtained by integrating by substitution where we set $y = \sec t$. Accordingly, other components $u_i(x) = 0$, for $i \geq 2$. Therefore, the exact solution

$$u(x) = \sec x \tan x, \tag{52}$$

follows immediately. It is clear that two components are calculated to determine the exact solution.

Exercises 3.2

Solve the following Volterra integral equations by using the *Adomian decomposition method*

1. $u(x) = 4x + 2x^2 - \int_0^x u(t)dt$

2. $u(x) = 1 + x - x^2 + \int_0^x u(t)dt$

3. $u(x) = 1 - \int_0^x u(t)dt$

4. $u(x) = x + \int_0^x (x-t)u(t)dt$

5. $u(x) = 3x - 9\int_0^x (x-t)u(t)dt$

6. $u(x) = 1 - 4\int_0^x (x-t)u(t)dt$

7. $u(x) = 1 + x - \int_0^x (x-t)u(t)dt$

8. $u(x) = 1 - x - \int_0^x (x-t)u(t)dt$

9. $u(x) = 1 + x + \int_0^x (x-t)u(t)dt$

10. $u(x) = 1 - x + \int_0^x (x-t)u(t)dt$

11. $u(x) = 2 + \int_0^x (x-t)u(t)dt$

12. $u(x) = 1 + x + \int_0^x u(t)dt$

13. $u(x) = 1 - \frac{1}{2!}x^2 - \int_0^x (x-t)u(t)dt$

14. $u(x) = 1 + \frac{1}{2!}x^2 + \int_0^x (x-t)u(t)dt$

In exercises 15 – 19 solve the given Volterra integral equations by using the *modified decomposition method*:

15. $u(x) = \cos x + \left(1 - e^{\sin x}\right)x + x\int_0^x e^{\sin t}u(t)dt$

16. $u(x) = \sec^2 x + \left(1 - e^{\tan x}\right)x + x\int_0^x e^{\tan t}u(t)dt,\ x < \pi/2$

17. $u(x) = \cosh x + \frac{x}{2}\left(1 - e^{\sinh x}\right) + \frac{x}{2}\int_0^x e^{\sinh t}u(t)dt$

18. $u(x) = \sinh x + \frac{1}{10}\left(e - e^{\cosh x}\right) + \frac{1}{10}\int_0^x e^{\cosh t}u(t)dt$

19. $u(x) = x^3 - x^5 + 5\int_0^x tu(t)dt$

20. $u(x) = \sec x \tan x + (e - e^{\sec x}) + \int_0^x e^{\sec t}u(t)dt,\ x < \pi/2$

3.3 The Series Solution Method

In this section we will introduce a practical method to solve the Volterra integral equation with variable limits of integration

$$u(x) = f(x) + \lambda\int_0^x K(x,t)u(t)dt, \tag{53}$$

where $K(x,t)$ is the kernel of the integral equation, and λ is a parameter. In the *series solution method* we will follow a parallel approach to the method of the series solution that usually applied in solving an ordinary differential equation around an ordinary point. The method is applicable provided that $u(x)$ is an analytic function, i.e. $u(x)$ has a Taylor expansion around $x = 0$. Accordingly, $u(x)$ can be expressed by a series expansion [24] given by

$$u(x) = \sum_{n=0}^{\infty} a_n x^n, \tag{54}$$

where the coefficients a_n are constants that will be determined. Substituting (54) into both sides of (53) yields

$$\sum_{n=0}^{\infty} a_n x^n = f(x) + \lambda \int_0^x K(x,t) \left(\sum_{n=0}^{\infty} a_n t^n \right) dt, \tag{55}$$

so that by using few terms of the expansions in both sides, we find

$$\begin{aligned} a_0 + a_1 x + a_2 x^2 + a_3 x^3 + \cdots \quad = f(x) \quad & +\lambda \textstyle\int_0^x K(x,t) a_0 \, dt \\ & +\lambda \int_0^x K(x,t) a_1 \, t dt \\ & +\lambda \int_0^x K(x,t) a_2 \, t^2 dt \\ & +\lambda \int_0^x K(x,t) a_3 \, t^3 dt \\ & + \cdots \end{aligned} \tag{56}$$

In view of (56), the integral equation (53) will be reduced to several traditional integrals, with defined integrands having terms of the form t^n, $n \geq 0$ only. We then write the Taylor expansions for $f(x)$ and evaluate the first few integrals in (56). Having performed the evaluation, we equate the coefficients of like powers of x in both sides of Eq. (56). This will lead to a complete determination of the coefficients $a_0, a_1, a_2, \cdots$. Consequently, substituting the obtained coefficients a_n, $n \geq 0$, in (54) produces the solution in a series form. This may lead to a solution in a closed form if the expansion obtained is a Taylor expansion to a well known elementary function.

It seems reasonable to illustrate the series solution method by discussing the following examples.

Example 1. We use the series solution method to solve

$$u(x) = 1 + \int_0^x (t - x) u(t) dt. \tag{57}$$

Substituting $u(x)$ by the series

$$u(x) = \sum_{n=0}^{\infty} a_n x^n, \tag{58}$$

into both sides of the equation (57) leads to

$$\sum_{n=0}^{\infty} a_n x^n = 1 + \int_0^x (t-x)\left(\sum_{n=0}^{\infty} a_n t^n\right) dt, \tag{59}$$

which gives

$$\sum_{n=0}^{\infty} a_n x^n = 1 + \int_0^x \left(\sum_{n=0}^{\infty} a_n t^{n+1} - x\sum_{n=0}^{\infty} a_n t^n\right) dt. \tag{60}$$

Evaluating the regular integrals at the right hand side that involve terms of the form t^n, $n \geq 0$ yields

$$\sum_{n=0}^{\infty} a_n x^n = 1 - \sum_{n=0}^{\infty} \frac{1}{(n+1)(n+2)} a_n x^{n+2}, \tag{61}$$

or equivalently

$$a_0 + a_1 x + a_2 x^2 + a_3 x^3 + \cdots = 1 - \frac{1}{2!}a_0 x^2 - \frac{1}{3!}a_1 x^3 - \frac{1}{12}a_2 x^4 + \cdots \tag{62}$$

Equating the coefficients of like powers of x in both sides we find

$$a_0 = 1, \tag{63}$$

$$a_1 = 0, \tag{64}$$

$$a_2 = -\frac{1}{2!}, \tag{65}$$

$$a_3 = 0, \tag{66}$$

$$a_4 = \frac{1}{4!}, \tag{67}$$

and generally

$$a_{2n} = (-1)^n \frac{1}{(2n)!}, \quad \text{for} \quad n \geq 0, \tag{68}$$

$$a_{2n+1} = 0, \quad \text{for} \quad n \geq 0. \tag{69}$$

Using (58) we find the solution in a series form

$$u(x) = 1 - \frac{1}{2!}x^2 + \frac{1}{4!}x^4 - \frac{1}{6!}x^6 + \ldots, \tag{70}$$

and in a closed form

$$u(x) = \cos x. \tag{71}$$

Example 2. We next use the series solution method to solve

$$u(x) = 2\cosh x - x\sinh x - 1 + \int_0^x tu(t)dt. \tag{72}$$

Proceeding as before, and substituting (58) into both sides of the equation (72) we obtain

$$\sum_{n=0}^{\infty} a_n x^n = 2\left(\sum_{n=0}^{\infty} \frac{x^{2n}}{(2n)!}\right) - x\left(\sum_{n=0}^{\infty} \frac{x^{2n+1}}{(2n+1)!}\right) - 1 + \int_0^x t\left(\sum_{n=0}^{\infty} a_n t^n\right) dt. \tag{73}$$

Using few terms from each series involved and integrating the resulting integrals at the right hand side we obtain

$$a_0 + a_1 x + a_2 x^2 + \cdots = 2\left(1 + \frac{x^2}{2!} + \frac{x^4}{4!} + \cdots\right) - x\left(x + \frac{x^3}{3!} + \frac{x^5}{5!}\cdots\right)$$

$$-1 + \frac{1}{2}a_0 x^2 + \frac{1}{3}a_1 x^3 + \frac{1}{4}a_2 x^4 + \cdots \tag{74}$$

Equating the coefficients of like powers of x in in both sides of (74) yields

$$a_0 = 1, \tag{75}$$

$$a_1 = 0, \tag{76}$$

$$a_2 = \frac{1}{2!}, \tag{77}$$

$$a_3 = 0, \tag{78}$$

$$a_4 = \frac{1}{4!}, \tag{79}$$

and generally

$$a_{2n} = \frac{1}{(2n)!}, \quad \text{for } n \geq 0, \tag{80}$$

$$a_{2n+1} = 0, \quad \text{for } n \geq 0. \tag{81}$$

Consequently, the solution in a series form is given by

$$u(x) = 1 + \frac{1}{2!}x^2 + \frac{1}{4!}x^4 + \frac{1}{6!}x^6 + \ldots, \tag{82}$$

and in a closed form

$$u(x) = \cosh x, \tag{83}$$

obtained by using the Taylor expansion of $\cosh x$.

Exercises 3.3

Solve the following Volterra integral equations by using the *series solution method*

1. $u(x) = 2x + 2x^2 - x^3 + \int_0^x u(t)dt$
2. $u(x) = 1 + x - \frac{2}{3}x^3 - \frac{1}{2}x^4 + 2\int_0^x tu(t)dt$
3. $u(x) = 1 + 2\sin x - \int_0^x u(t)dt$
4. $u(x) = 1 + x + \frac{1}{2!}x^2 + \frac{1}{3!}x^3 - \int_0^x (x-t)u(t)dt$
5. $u(x) = -1 - \int_0^x u(t)dt$
6. $u(x) = 1 - 2\int_0^x u(t)dt$
7. $u(x) = 1 + xe^x - \int_0^x tu(t)dt$
8. $u(x) = x + \int_0^x (x-t)u(t)dt$
9. $u(x) = 1 - \frac{1}{2!}x^2 - \int_0^x (x-t)u(t)dt$
10. $u(x) = 1 - x - \int_0^x (x-t)u(t)dt$
11. $u(x) = 1 + \sinh x - \cosh x + \int_0^x u(t)dt$
12. $u(x) = x\cos x + \int_0^x tu(t)dt$

3.4 Converting Volterra Equation to IVP

In Chapter 1 the process of converting initial value problems to equivalent Volterra integral equations has been discussed in details. However, the technique of reducing Volterra integral equations to initial value problems will be discussed in this section, though it is rarely used. This may be explained by the fact that integral equations are easily solved because initial conditions are embedded in the integral equations. However, solving initial value problems ,where initial conditions will be used, will increase the size of evaluations required because additional steps will be needed to complete the solution.

To employ this method, we simply differentiate both sides of (1), noting that Leibnitz rule should be used in differentiating the integral at the right hand side of (1). The differentiation procedure should be carried successively until the integral sign is removed and the integral equation is reduced to a pure differential equation equivalent to the integral equation under discussion. It is interesting to note that initial conditions should be determined at every step of differentiating by setting $x = 0$ at $u(x)$ and its obtained derivatives. The resulting initial value problem is then solved following the traditional techniques used in undergraduate course of ordinary differential equations. The technique of converting Volterra integral equations to initial value problems, though not usually used as indicated earlier, will be illustrated by discussing the following examples.

Example 1. Solve the following Volterra integral equation

$$u(x) = x^2 + \frac{1}{12}x^4 + \int_0^x (t - x)\, u(t)\, dt, \tag{84}$$

by reducing to an equivalent initial value problem.

Differentiating both sides of (84) with respect to x and using the Leibnitz rule yield

$$u^{'}(x) = 2x + \frac{1}{3}x^3 - \int_0^x u(t)\, dt. \tag{85}$$

Note that we have to differentiate both sides of (85) again to get rid of the integral sign at the right hand side, doing so we obtain

$$u^{''}(x) = 2 + x^2 - u(x), \tag{86}$$

or equivalently the nonhomogeneous second order differential equation

$$u''(x) + u(x) = 2 + x^2. \tag{87}$$

The proper initial conditions can be obtained by substituting $x = 0$ into both sides of equations (84) and (85), hence we find

$$u(0) = 0, \quad u'(0) = 0. \tag{88}$$

To solve the resulting initial value problem

$$u''(x) + u(x) = 2 + x^2, \quad u(0) = 0, \quad u'(0) = 0, \tag{89}$$

we first solve the corresponding homogeneous equation

$$u''(x) + u(x) = 0. \tag{90}$$

The characteristic equation of (90) is given by

$$r^2 + 1 = 0, \tag{91}$$

so that the roots of (91) are

$$r_1 = i, \quad r_2 = -i. \tag{92}$$

As a result the complementary solution is

$$u_c = A\cos x + B\sin x, \tag{93}$$

where A and B are constants to be determined later by using the initial conditions. To determine a particular solution $u_p(x)$ of (89), we assume that $u_p(x)$ is of the form

$$u_p = \alpha + \beta x + \gamma x^2, \tag{94}$$

where α , β , and γ are constants that will be determined. Substituting (94) into (89), and equating like powers of x on each side we find

$$\alpha = 0, \quad \beta = 0, \quad \gamma = 1. \tag{95}$$

Combining (93) – (95) we obtain the general solution

$$u(x) = A\cos x + B\sin x + x^2. \tag{96}$$

We can determine the constants A and B upon using the initial conditions (88) where we find

$$A = 0, \quad B = 0, \tag{97}$$

so that the solution of (89)

$$u(x) = x^2, \tag{98}$$

follows immediately.

However, we can easily show that using the modified decomposition method, by setting $u_0(x) = x^2$, will give the same result obtained above with minimal work.

Example 2. Solve the following Volterra integral equation

$$u(x) = e^x + \int_0^x (t - x)\, u(t)\, dt, \tag{99}$$

by converting to an equivalent initial value problem.

Differentiating both sides of (99) twice with respect to x and using the Leibnitz rule yield

$$u'(x) = e^x - \int_0^x u(t)\, dt, \tag{100}$$

and

$$u''(x) = e^x - u(x) \tag{101}$$

or equivalently

$$u''(x) + u(x) = e^x. \tag{102}$$

The proper initial conditions can be obtained by substituting $x = 0$ into both sides of the equations (99) and (100), hence we find

$$u(0) = 1, \quad u'(0) = 1. \tag{103}$$

To solve the resulting initial value problem

$$u''(x) + u(x) = e^x, \quad u(0) = 1, \quad u'(0) = 1, \tag{104}$$

we first solve the corresponding homogeneous equation

$$u''(x) + u(x) = 0. \tag{105}$$

Following the steps we used in the first example, we find that the complementary solution is

$$u_c = A\cos x + B\sin x, \tag{106}$$

where A and B are constants to be determined later by using the initial conditions. Moreover, a particular solution of (104) may be assumed of the form

$$u_p = \alpha e^x, \tag{107}$$

where α is a constant that will be determined. Substituting (107) into (104) and equating like powers of x on both sides we find

$$\alpha = \frac{1}{2}. \tag{108}$$

Combining (106) – (108) we obtain the general solution

$$u(x) = A\cos x + B\sin x + \frac{1}{2}e^x. \tag{109}$$

The constants A and B are determined by using the initial conditions in (103) where we find

$$A = \frac{1}{2}, \quad B = \frac{1}{2}. \tag{110}$$

Thus the solution of (104) is

$$u(x) = \frac{1}{2}\left(\sin x + \cos x + e^x\right), \tag{111}$$

obtained upon substituting (110) into (109) .

Exercises 3.4

Solve the following Volterra integral equations by converting to equivalent *initial value problems:*

1. $u(x) = 1 - 3\int_0^x u(t)dt$

2. $u(x) = 1 + \int_0^x (x-t)u(t)dt$

3. $u(x) = 1 - x - \int_0^x (x-t)u(t)dt$

4. $u(x) = x + \int_0^x u(t)dt$

5. $u(x) = 1 + x + \int_0^x (x-t)u(t)dt$

6. $u(x) = 1 + \frac{1}{6}\int_0^x (x-t)^3 u(t)dt$

7. $u(x) = x + \frac{1}{6}\int_0^x (x-t)^3 u(t)dt$

8. $u(x) = x^2 + \int_0^x (x-t)u(t)dt$

9. $u(x) = x + \frac{1}{3!}x^3 - \int_0^x (x-t)u(t)dt$

10. $u(x) = x - x^2 + \frac{1}{6}x^3 - \frac{1}{12}x^4 - \int_0^x (x-t)u(t)dt$

3.5 Successive Approximations Method

The method of successive approximations used in Section 2.4 for handling Fredholm integral equation will be implemented here to solve Volterra integral equation. In this method, we replace the unknown function $u(x)$ under the integral sign of the Volterra equation

$$u(x) = f(x) + \lambda \int_0^x K(x,t)u(t)dt, \qquad (112)$$

by any selective real valued continuous function $u_0(x)$, called the zeroth approximation. This substitution will give the first approximation $u_1(x)$ by

$$u_1(x) = f(x) + \lambda \int_0^x K(x,t)u_0(t)dt. \qquad (113)$$

It is obvious that $u_1(x)$ is continuous whenever $f(x)$, $K(x,t)$ and $u_0(x)$ are continuous. The second approximation $u_2(x)$ of $u(x)$ can be obtained similarly by replacing $u_0(x)$ in (113) by $u_1(x)$ obtained above, hence we find

$$u_2(x) = f(x) + \lambda \int_0^x K(x,t)u_1(t)dt. \qquad (114)$$

This process can be continued in the same manner to obtain the *nth* approximation. In other words, the various approximations of the solution $u(x)$ of (112) can be obtained in a recursive scheme given by

$$\begin{cases} u_0(x) = \text{any selective real valued function} \\ u_n(x) = f(x) + \lambda \int_0^x K(x,t)u_{n-1}(t)dt, \quad n \geq 1. \end{cases} \tag{115}$$

The most commonly selected functions for $u_0(x)$ are 0, 1 or x. At the limit, the solution $u(x)$ of the equation (112) is obtained by

$$u(x) = \lim_{n \to \infty} u_n(x), \tag{116}$$

so that the resulting solution $u(x)$ is independent of the choice of the zeroth approximation $u_0(x)$.

It is useful, for comparison reasons, to distinguish between the recursive schemes used in the decomposition method and in the successive approximations method. In the decomposition method, we decompose the solution $u(x)$ into components $u_0, u_1, u_2, \ldots$ where each component is evaluated subsequently, and in this case the solution is given in a series form

$$u(x) = \sum_{n=0}^{\infty} u_n(x), \tag{117}$$

where the zeroth component $u_0(x)$ is defined and given by all terms that are out of the integral sign. However, in the successive approximations method, we apply the above recursive scheme (115) to determine various approximations of the solution $u(x)$ itself, and not components of $u(x)$, noting that the zeroth approximation is not defined but rather given by a selective real valued function, and as a result the solution $u(x)$ is given by the formula (116) .

To illustrate the difference between the two recursive algorithms , we start by solving Example 2 in Section 3.2. The reader can easily compare between the two approaches.

Example 1. Solve the Volterra integral equation

$$u(x) = x + \int_0^x (t - x)u(t)\, dt, \tag{118}$$

by the *successive approximations method.* We first select any real valued function for the zeroth approximation, hence we set

$$u_0(x) = 0. \tag{119}$$

Substituting (119) into (118) we find

$$u_1(x) = x + \int_0^x (t - x)\, u_0(t)\, dt, \tag{120}$$

and this gives the first approximation of $u(x)$ by

$$u_1(x) = x. \tag{121}$$

Inserting (121) into (120) to replace $u_0(x)$ we obtain

$$u_2(x) = x + \int_0^x (t - x)\, t\, dt, \tag{122}$$

where by integrating the right hand side of (122), the second approximation of $u(x)$

$$u_2(x) = x - \frac{1}{3!}x^3, \tag{123}$$

is readily obtained. Continuing in the same manner we find that the third approximation of $u(x)$ is

$$u_3(x) = x - \frac{1}{3!}x^3 + \frac{1}{5!}x^5. \tag{124}$$

Accordingly, the nth approximation is given by

$$u_n(x) = \sum_{k=1}^{n} (-1)^{k-1} \frac{x^{2k-1}}{(2k-1)!}, \quad n \geq 1. \tag{125}$$

Consequently, the solution $u(x)$ of (118) is given by

$$\begin{aligned} u(x) &= \lim_{n\to\infty} u_n(x), \\ &= \lim_{n\to\infty} \left(\textstyle\sum_{k=1}^{n} (-1)^{k-1} \frac{x^{2k-1}}{(2k-1)!} \right) \qquad (126) \\ &= \sin x. \end{aligned}$$

To show that $u(x)$ obtained in (126) does not depend on the selection of $u_0(x)$, we will solve the equation (118) by selecting

$$u_0(x) = x. \tag{127}$$

Using the new selection of $u_0(x)$ in the right hand side of (118) we obtain

$$u_1(x) = x + \int_0^x (t - x)\, t dt, \tag{128}$$

which gives the first approximation by

$$u_1(x) = x - \frac{1}{3!}x^3. \tag{129}$$

Proceeding as before we can easily obtain the second approximation

$$u_2(x) = x - \frac{1}{3!}x^3 + \frac{1}{5!}x^5. \tag{130}$$

In a parallel manner we find that

$$u_n(x) = \sum_{k=0}^{n} (-1)^k \frac{x^{2k+1}}{(2k+1)!}, \quad n \geq 0. \tag{131}$$

Accordingly, we obtain

$$\begin{aligned} u(x) &= \lim_{n\to\infty} u_n(x), \\ &= \lim_{n\to\infty} \left(\textstyle\sum_{k=0}^{n} (-1)^k \frac{x^{2k+1}}{(2k+1)!}\right) \\ &= \sin x, \end{aligned} \tag{132}$$

the same answer we obtained above in (126). This confirms the fact that the solution obtained does not depend on the selection of the zeroth approximation $u_0(x)$.

Example 2. Solve the Volterra integral equation

$$u(x) = 1 - \int_0^x (t - x)\, u(t) dt, \tag{133}$$

by using the successive approximations method. We start first by selecting the zeroth approximation and in this time we choose

$$u_0(x) = 1, \tag{134}$$

where by substituting this in the right hand side of (133) the first approximation

$$u_1(x) = 1 - \int_0^x (t-x)u_0(t)\,dt, \tag{135}$$

so that

$$u_1(x) = 1 + \frac{1}{2!}x^2 \tag{136}$$

follows immediately. Proceeding in the same manner we find that

$$u_2(x) = 1 - \int_0^x (t-x)\,u_1(t)dt, \tag{137}$$

so that

$$u_2(x) = 1 + \frac{1}{2!}x^2 + \frac{1}{4!}x^4. \tag{138}$$

In a similar manner we obtain

$$u_3(x) = 1 + \frac{1}{2!}x^2 + \frac{1}{4!}x^4 + \frac{1}{6!}x^4. \tag{139}$$

Generally we obtain for the nth approximation

$$u_n(x) = \sum_{k=0}^{n} \frac{x^{2k}}{(2k)!},\ n \geq 0. \tag{140}$$

Consequently, the solution $u(x)$ of (133) is given by

$$\begin{aligned} u(x) &= \lim_{n\to\infty} u_n(x), \\ &= \lim_{n\to\infty}\left(\sum_{k=0}^{n} \frac{x^{2k}}{(2k)!}\right) \\ &= \cosh x, \end{aligned} \tag{141}$$

obtained upon using the Taylor expansion of $\cosh x$.

It is useful to observe that the zeroth approximation in this method is selected and it is not a part of the integral equation as in the decomposition method.

Exercises 3.5

Solve the following Volterra integral equations by the *successive approximations method:*

1. $u(x) = 1 - \int_0^x u(t)dt$

2. $u(x) = 1 - 9\int_0^x (x-t)u(t)dt$

3. $u(x) = 1 + 2x + 4\int_0^x (x-t)u(t)dt$

4. $u(x) = 1 - \frac{1}{4}x + \frac{1}{16}\int_0^x (x-t)u(t)dt$

5. $u(x) = 2 - \int_0^x (x-t)u(t)dt$

6. $u(x) = 1 - \int_0^x 2tu(t)dt$

7. $u(x) = x + \int_0^x (x-t)u(t)dt$

8. $u(x) = 1 - \int_0^x (x-t)u(t)dt$

9. $u(x) = 1 + x - \int_0^x (x-t)u(t)dt$

10. $u(x) = 1 - x - \int_0^x (x-t)u(t)dt$

11. $u(x) = 2 - x + \int_0^x u(t)dt$

12. $u(x) = 1 - x - \frac{1}{2}x^2 + \int_0^x (x-t)u(t)dt$

3.6 The Method of Successive Substitutions

The technique to be used here is completely identical to that used in Section 2.5. In this method, we set $x = t$ and $t = t_1$ in the Volterra integral equation

$$u(x) = f(x) + \lambda\int_0^x K(x,t)u(t)dt, \tag{142}$$

to obtain

$$u(t) = f(t) + \lambda \int_0^t K(t, t_1)u(t_1)dt_1. \tag{143}$$

Replacing $u(t)$ at the right hand side of (142) by its obtained value given by (143) yields

$$\begin{aligned} u(x) \quad = f(x) \quad &+\lambda \int_0^x K(x,t)f(t)dt \\ &+\lambda^2 \int_0^x K(x,t) \int_0^t K(t,t_1)u(t_1)dt_1dt. \end{aligned} \tag{144}$$

Substituting $x = t_1$ and $t = t_2$ in (142) we obtain

$$u(t_1) = f(t_1) + \lambda \int_0^{t_1} K(t_1, t_2)u(t_2)dt_2. \tag{145}$$

Substituting the value of $u(t_1)$ obtained in (145) into the right hand side of (144) leads to

$$\begin{aligned} u(x) \quad = f(x) \quad &+\lambda \int_0^x K(x,t)f(t)dt \\ &+\lambda^2 \int_0^x \int_0^t K(x,t)K(t,t_1)f(t_1)dt_1dt \\ &+\lambda^3 \int_0^x \int_0^t \int_0^{t_1} K(x,t)K(t,t_1)K(t_1,t_2)u(t_2)dt_2dt_1dt. \end{aligned} \tag{146}$$

Accordingly, the general series form for $u(x)$ can be rewritten as

$$\begin{aligned} u(x) \quad = \quad & f(x) + \lambda \int_0^x K(x,t)f(t)dt \\ &+\lambda^2 \int_0^x \int_0^t K(x,t)K(t,t_1)f(t_1)dt_1dt \\ &+\lambda^3 \int_0^x \int_0^t \int_0^{t_1} K(x,t)K(t,t_1)K(t_1,t_2)f(t_2)dt_2dt_1dt \\ &+\cdots \end{aligned} \tag{147}$$

We remark here that in this method the unknown function $u(x)$ is substituted by the given function $f(x)$ that makes the evaluation of the multiple integrals easily computable. This substitution of $u(x)$ occurs several times through the integrals and hence this is why it is called the method of successive substitutions. The technique will be illustrated by discussing the following examples.

Example 1. We solve the following Volterra integral equation

$$u(x) = x - \int_0^x (x-t)u(t)dt, \tag{148}$$

by using the method of successive substitutions. Substituting $\lambda = -1$, $f(x) = x$, and $K(x,t) = (x-t)$ into (147) yields

$$u(x) = x - \int_0^x (x-t)\, tdt + \int_0^x \int_0^t (x-t)(t-t_1)t_1 dt_1 dt + \cdots, \tag{149}$$

or equivalently

$$u(x) = x - \int_0^x (xt - t^2)dt + \int_0^x \int_0^t (x-t)(tt_1 - t_1^2)dt_1 dt + \cdots. \tag{150}$$

Therefore we obtain the solution in a series form

$$u(x) = x - \frac{1}{3!}x^3 + \frac{1}{5!}x^5 + \cdots, \tag{151}$$

or in a closed form

$$u(x) = \sin x \tag{152}$$

upon using the Taylor expansion for $\sin x$.

Example 2. We next solve the Volterra integral equation by applying the method of successive substitutions

$$u(x) = x^2 - x^4 + \int_0^x 4t\, u(t)dt. \tag{153}$$

Substituting $\lambda = 4$, $f(x) = x^2 - x^4$, and $K(x,t) = t$ into (147) yields

$$\begin{aligned} u(x) &= x^2 - x^4 + 4\int_0^x t\left(t^2 - t^4\right)dt + 16\int_0^x \int_0^t tt_1\left(t_1^2 - t_1^4\right)dt_1 dt \\ &\quad +64\int_0^x \int_0^t \int_0^{t_1} tt_1t_2\left(t_2^2 - t_2^4\right)dt_2 dt_1 dt \\ &\quad +\cdots, \end{aligned} \tag{154}$$

and this will yield

$$u(x) = x^2 - x^4 + x^4 - \frac{2}{3}x^6 + \frac{2}{3}x^6 - \frac{1}{3}x^8 + \frac{1}{3}x^8 + \cdots. \qquad (155)$$

Consequently, we easily obtain the exact solution

$$u(x) = x^2, \qquad (156)$$

upon cancelling the similar terms with opposite signs in (155).

Exercises 3.6

Solve the following Volterra integral equations by the *successive substitutions method:*

1. $u(x) = x + \int_0^x u(t)dt$

2. $u(x) = \frac{1}{2!}x^2 + \int_0^x u(t)dt$

3. $u(x) = \frac{1}{3!}x^3 - \int_0^x (x-t)u(t)dt$

4. $u(x) = \frac{1}{3!}x^3 + \int_0^x (x-t)u(t)dt$

5. $u(x) = \frac{1}{2!}x^2 - \int_0^x (x-t)u(t)dt$

6. $u(x) = 1 - \frac{1}{2!}x^2 - \int_0^x u(t)dt$

7. $u(x) = 1 + 2\int_0^x u(t)dt$

8. $u(x) = 3 - 2x + \int_0^x u(t)dt$

9. $u(x) = 2 + \frac{1}{2}x^2 - \int_0^x (x-t)u(t)dt$

10. $u(x) = 1 - x + \frac{1}{2}x^2 - \int_0^x (x-t)u(t)dt$

11. $u(x) = 2 - \frac{1}{2}x^2 + \int_0^x (x-t)u(t)dt$

12. $u(x) = \frac{1}{2!}x^2 + \int_0^x (x-t)u(t)dt$

3.7 Comparison between Alternative Methods

Before making a comparison between all methods discussed, it is convenient to point out that there are other techniques for solving Volterra integral equations that are beyond the scope of this text. Using Laplace transforms to handle Volterra equations requires extensive background in Laplace transforms and the convolution integral. However, when it comes to select a preferable method among the methods that were introduced in the previous sections, we cannot recommend a specific method.

Even though the method of reducing Volterra integral equation to initial value problem is rarely used, but it is the only method that may give directly the exact solution in a closed form. This is easily seen if the resulting initial value problem has constant coefficients. We may obtain the solution in a series form, similar to the results obtained by other methods, when the coefficients of the resulting initial value problem are functions of the independent variable x. The latter case will not be introduced here. A useful example has been discussed by Example 3 in Section 3.2.

However, we found that if the kernel $K(x,t)$ of the integral equation is a degenerate one that consists of a polynomial of one or two terms, the series method and the decomposition method might be the best choices because it minimize the volume of calculations. The series solution obtained by using these methods might yield the exact solution in a closed form or we may obtain an approximation of the solution. Moreover, if $f(x)$ is a transcendental function, the series solution method works easier than the decomposition method.

Comparing the Adomian decomposition method with the successive approximation method, it is evident the the decomposition method is much easier in that we integrate always very few terms to obtain the successive components, whereas in the other method we integrate many terms to evaluate the successive approximations after selecting the zeroth approximation.

In closing this section, we point out that the method of successive substitutions suffers from the huge size of calculations, especially if the function $f(x)$ is a trigonometric or exponential function.

To achieve our goal of the comparison between these methods, we illustrate this comparison by solving the following Volterra integral equation by using all various methods.

Example 1. We solve the following example

$$u(x) = 1 + \int_0^x u(t)dt, \tag{157}$$

by using the five alternative methods.

(a) Adomian Decomposition Method: As discussed earlier we set

$$u(x) = \sum_{n=0}^{\infty} u_n(x). \tag{158}$$

Substituting (158) into both sides of (157) we obtain

$$u_0(x) + u_1(x) + u_2(x) + \cdots = 1 + \int_0^x (u_0(t) + u_1(t) + u_2(t) + \cdots)\, dt. \tag{159}$$

As stated before, we have to set the zeroth component $u_0(x)$ by all terms outside the integral sign, hence we have

$$u_0(x) = 1. \tag{160}$$

The first component $u_1(x)$ may be obtained by

$$u_1(x) = \int_0^x u_0(t)dt, \tag{161}$$

which gives the first component

$$u_1(x) = x. \tag{162}$$

Proceeding in the same manner we can easily obtain

$$u_2(x) = \frac{1}{2!}x^2, \tag{163}$$

and so on for other components. Noting that in the decomposition method we have

$$u(x) = u_0 + u_1 + u_2 + u_3 + \cdots, \tag{164}$$

hence by using the results (160) – (163) into (164) we obtain the solution in a series form

$$u(x) = 1 + x + \frac{1}{2!}x^2 + \frac{1}{3!}x^3 + \cdots, \tag{165}$$

and in a closed form the exact solution

$$u(x) = e^x, \tag{166}$$

follows immediately.

(b) The Series Method: As indicated before, assuming that $u(x)$ is analytic, hence we may write

$$u(x) = \sum_{n=0}^{\infty} a_n\, x^n. \tag{167}$$

Substituting (167) into both sides of (157) we find

$$a_0 + a_1x + a_2x^2 + a_3x^3 + \cdots = 1 + \int_0^x \left(a_0 + a_1t + a_2t^2 + \cdots\right) dt. \tag{168}$$

Integrating the easy integrals in the right hand side we obtain

$$a_0 + a_1x + a_2x^2 + a_3x^3 + \cdots = 1 + a_0x + \frac{1}{2}a_1x^2 + \frac{1}{3}a_2x^3 + \cdots. \tag{169}$$

Equating the coefficients of like powers of x from in sides we find

$$a_0 = 1, \tag{170}$$

$$a_1 = 1, \tag{171}$$

$$a_2 = \frac{1}{2!}, \tag{172}$$

$$a_3 = \frac{1}{3!}, \tag{173}$$

and so on. Substituting the results obtained for a_k, $k \geq 0$ into (167) we obtain the solution in a series form

$$u(x) = \sum_{n=0}^{\infty} \frac{x^n}{n!}, \tag{174}$$

and in a closed form

$$u(x) = e^x, \tag{175}$$

is the exact solution of the example under discussion.

(c) Converting to Initial Value Problems: Differentiating both sides of (157) with respect to x we obtain

$$u'(x) = u(x), \tag{176}$$

with the initial condition

$$u(0) = 1. \tag{177}$$

Solving the first separable order differential equation (176) and using the initial condition (177) the exact solution

$$u(x) = e^x, \tag{178}$$

is readily obtained.

(d) Successive Approximations Method: In this method we select the zeroth approximation by

$$u_0(x) = 1. \tag{179}$$

Following the technique that was discussed above, the other approximations of the solution $u(x)$ can be easily obtained by

$$u_1(x) = 1 + x, \tag{180}$$

$$u_2(x) = 1 + x + \frac{x^2}{2!}, \tag{181}$$

$$u_3(x) = 1 + x + \frac{x^2}{2!} + \frac{x^3}{3!}, \tag{182}$$

and so on. Accordingly, the nth component is given by

$$u_n(x) = 1 + x + \frac{x^2}{2!} + \frac{x^3}{3!} + \cdots + \frac{x^n}{n!}. \tag{183}$$

Consequently we find

$$\begin{aligned} u(x) &= \lim_{n\to\infty} u_n(x), \\ &= \lim_{n\to\infty} 1 + x + \tfrac{x^2}{2!} + \tfrac{x^3}{3!} + \cdots \\ &= e^x, \end{aligned} \tag{184}$$

the same solution obtained above.

(e) The Method of Successive Substitutions: In this method we have to set $K(x,t) = 1$, $\lambda = 1$ and $f(x) = 1$, hence we have

$$\begin{aligned} u(x) &= 1 + 1\int_0^x 1dt + \int_0^x\int_0^t 1dt_1dt + \int_0^x\int_0^t\int_0^{t_1} 1dt_2dt_1dt + \cdots \\ &= 1 + x + \frac{x^2}{2!} + \frac{x^3}{3!} + \cdots \\ &= e^x, \end{aligned} \tag{185}$$

the same result obtained by other methods.

An important observation, and not a recommendation,can be made from the comparison performed above which suggests that the decomposition method and the series method introduce promising improvements over other existing techniques.

3.8 Volterra Equations of the First Kind

In this section we will study the Volterra integral equation of the first kind with separable kernel given by

$$f(x) = \int_0^x K(x,t)\, u(t)dt. \tag{186}$$

It is important to note that the Volterra integral equation of the first kind can be handled simply by reducing this equation to Volterra equation of the second kind. This goal can be accomplished by differentiating both sides of (186) with respect to x to obtain

$$f'(x) = K(x,x)u(x) + \int_0^x K_x(x,t)u(t)dt, \tag{187}$$

by using Leibnitz rule. If $K(x,x) \neq 0$ in the interval of discussion, then dividing both sides of (187) by $K(x,x)$ yields

$$u(x) = \frac{f'(x)}{K(x,x)} - \frac{1}{K(x,x)} \int_0^x K_x(x,t)\, u(t)dt, \tag{188}$$

a Volterra integral equation of the second kind. The case where the kernel $K(x,x) = 0$ leads to a complicated behavior of the problem that will not be investigated here.

To solve (188) we select any method that we discussed before. The technique of differentiating both sides of Volterra integral equation of the first kind, verifying that $K(x,x) \neq 0$, reducing to Volterra integral equation of the second kind and solving the resulting equation will be illustrated by discussing the following examples.

Example 1. Find the solution of the Volterra equation of the first kind

$$x^2 + \frac{1}{6}x^3 = \int_0^x (2 + x - t)\, u(t)dt. \tag{189}$$

We note first that $K(x,t) = 2 + x - t$, hence $K(x,x) = 2 \neq 0$. Differentiating both sides of (189) with respect to x yields

$$2x + \frac{1}{2}x^2 = 2u(x) + \int_0^x u(t)dt, \tag{190}$$

or equivalently

$$u(x) = x + \frac{1}{4}x^2 - \frac{1}{2}\int_0^x u(t)dt. \tag{191}$$

We prefer to use the *modified decomposition method*, hence we set

$$u_0(x) = x, \tag{192}$$

which gives

$$u_1(x) = \frac{1}{4}x^2 - \frac{1}{2}\int_0^x t\,dt, \\ = 0. \qquad (193)$$

Accordingly, other components $u_n(x) = 0,\ n \geq 2$. The exact solution $u(x)$ is

$$u(x) = x, \qquad (194)$$

obtained upon using the components obtained above.

Example 2. Find the solution of the Volterra equation of the first kind

$$xe^x = \int_0^x e^{x-t}\,u(t)dt. \qquad (195)$$

We note first that $K(x,t) = e^{x-t}$, hence $K(x,x) = 1 \neq 0$. Differentiating both sides of (195) with respect to x yields

$$e^x + xe^x = u(x) + \int_0^x e^{x-t}u(t)dt, \qquad (196)$$

or equivalently

$$u(x) = e^x + xe^x - \int_0^x e^{x-t}u(t)\,dt. \qquad (197)$$

We shall solve the resulting equation by the *Adomian decomposition method* and by the *modified decomposition method.* We first start by applying the Adomian decomposition method, therefore we set

$$u_0(x) = e^x + xe^x. \qquad (198)$$

Consequently, we obtain

$$u_1(x) = -\int_0^x e^{x-t}\left(e^t + te^t\right)dt \\ = -e^x\left(x + \frac{x^2}{2!}\right), \qquad (199)$$

and

$$u_2(x) = \int_0^x e^{x-t}\left(te^t + \frac{t^2}{2!}e^t\right)dt \\ = e^x\left(\frac{x^2}{2!} + \frac{x^3}{3!}\right), \qquad (200)$$

and so on. Using the above results of the components obtained gives

$$\begin{aligned} u(x) &= e^x\left(1 + x - x - \frac{x^2}{2!} + \frac{x^2}{2!} + \frac{x^3}{3!} - \frac{x^3}{3!} - \cdots\right) dt \\ &= e^x, \end{aligned} \qquad (201)$$

the exact solution obtained upon cancelling like terms with opposite signs.

Using the *modified decomposition method* we set

$$u_0(x) = e^x, \qquad (202)$$

which gives

$$\begin{aligned} u_1(x) &= xe^x - \int_0^x e^x\, dt, \\ &= 0. \end{aligned} \qquad (203)$$

It immediately follows that

$$u_n(x) = 0, \;\; n \geq 1. \qquad (204)$$

Accordingly, the exact solution is

$$u(x) = e^x. \qquad (205)$$

Exercises 3.8

Solve the following Volterra integral equations of the first kind

1. $5x^2 + x^3 = \int_0^x (5 + 3x - 3t)u(t)dt.$

2. $xe^{-x} = \int_0^x e^{t-x}u(t)dt.$

3. $2e^x - x - 2 = \int_0^x (1 + x - t)u(t)dt.$

4. $2\cosh x - \sinh x - (2 - x) = \int_0^x (2 - x + t)u(t)dt.$

5. $4\sin x - 3\cos x + 3 = \int_0^x (4 + 3x - 3t)u(t)dt\ .$

6. $\tan x - \ln(\cos x) = \int_0^x (1 + x - t)u(t)dt, \quad x < \pi/2.$

Chapter 4

Integro-Differential Equations

4.1 Introduction

In this chapter we shall be concerned with the integro-differential equations where both differential and integral operators will appear in the same equation. This type of equations was introduced by Volterra [5], [6] and [22] for the first time in the early 1900. Volterra was investigating the population growth, focusing his study on the hereditary influences, where through his research work the topic of integro-differential equations was established.

Scientists and researchers investigated the topic of integro-differential equations through their work in science applications such as heat transfer, diffusion process in general, neutron diffusion and biological species coexisting together with increasing and decreasing rates of generating [14]. More details about the sources where these equations arise can be found in physics, biology and engineering applications as well as in advanced integral equations books such as [11], [14], [15] and [18].

In the integro-differential equations,it is important to note that the unknown function $u(x)$ and one or more of its derivatives such as $u'(x)$, $u''(x)$,... appear out and under the integral sign as well. One quick source of integro-differential equations can be clearly seen when we convert a differential equation to an integral equation using Leibnitz rule.

The integro-differential equation can be viewed in this case as an intermediate stage when finding an equivalent Volterra integral equation to the given differential equation as discussed in Section 1.5.

The following are examples of linear integro-differential equations:

$$
\begin{aligned}
u'(x) &= x - \int_0^1 e^{x-t} u(t)dt, \quad u(0) = 0, & (1)\\
u''(x) &= e^x - x + \int_0^1 xtu'(t)dt, \quad u(0) = 1, u'(0) = 1, & (2)\\
u'(x) &= x - \int_0^x (x-t)u(t)dt, \quad u(0) = 0, & (3)\\
u''(x) &= -x + \int_0^x (x-t)u(t)dt, \quad u(0) = 0, u'(0) = -1. & (4)
\end{aligned}
$$

It is clear from the examples given above that the unknown function $u(x)$ or one of its derivatives appear under the integral sign, and other derivatives of $u(x)$ appear out of the integral sign as well. Therefore, the above given equations involve the derivatives and the integral operators in the same equation, and consequently the term integro-differential equations has been used for problems involving this combination of operators.

Examining the limits of integrals in the equations (1) – (4) and following the classification concept used in Chapter 1 allow us to use the classification *Fredholm integro-differential equations* to the equations (1) and (2), and *Volterra integro-differential equations* to the equations (3) and (4). In addition, it is also interesting to know that the equations (1) – (4) are *linear* integro-differential equations, and this is related to the linearity occurrence of the unknown function $u(x)$ and its derivatives in the equations above. However, *nonlinear* integro-differential equations also arise in many scientific and engineering problems. Our concern in this chapter will be focused on the *linear* integro-differential equations by introducing several methods that handle this type of equations.

To determine a solution for the integro-differential equation, the initial conditions should be given, and this may be clearly seen as a result of involving $u(x)$ and its derivatives. The initial conditions are needed to determine the constants of integration.

In the following we will discuss several methods that handle successfully the linear integro-differential equations .

4.2 Fredholm Integro-Differential Equations

In this section we will discuss the reliable methods used to solve Fredholm integro-differential equations. We remark here that we will focus our concern on the equations that involve separable kernels where the kernel $K(x,t)$ can be expressed as a finite sum of the form

$$K(x,t)=\sum_{k=1}^{n} g_k(x)\,h_k(t). \tag{5}$$

Without loss of generality, we will make our analysis on a one term kernel $K(x,t)$ of the form

$$K(x,t)=g(x)\,h(t), \tag{6}$$

and this can be generalized for other cases. The nonseparable kernel can be reduced to separable kernel by using the Taylor expansion for the kernel involved. It is worthnoting that in this section we will introduce the most recent and practical schemes that handle this type of equations, where we may obtain an exact solution or an approximation to the solution with the highest desirable accuracy. We point out here that the methods to be discussed are introduced before, but we will focus our discussion on how these methods can be implemented in this type of equations. We first start with the most practical method.

4.2.1 The Direct Computation Method:

This method has been extensively introduced in Chapter 2. Without loss of generality, we may assume a standard form to the Fredholm integro-differential equation given by

$$u^{(n)}(x)=f(x)+\int_0^1 K(x,t)\,u(t)dt, \quad u^{(k)}(0)=b_k,\ 0\le k\le(n-1), \tag{7}$$

where $u^{(n)}(x)$ indicates the nth derivative of $u(x)$ with respect to x and b_k are constants that define the proper initial conditions. Substituting (6) into (7) yields

$$u^{(n)}(x) = f(x) + g(x) \int_0^1 h(t)\, u(t) dt, \quad u^{(k)}(0) = b_k,\ 0 \leq k \leq (n-1). \tag{8}$$

We can easily observe that the definite integral in the integro-differential equation (8) involves an integrand that completely depends on the variable t, and therefore, it seems reasonable to set that definite integral in the right hand side of (8) to a constant α, i.e. we set

$$\alpha = \int_0^1 h(t)u(t)\, dt. \tag{9}$$

With α defined in (9), the equation (8) can be written by

$$u^{(n)}(x) = f(x) + \alpha\, g(x). \tag{10}$$

It remains to determine the constant α to evaluate the exact solution $u(x)$. To find α, we should derive a form for $u(x)$ by using (10), followed by substituting this form in (9). To achieve this we integrate both sides of (10) n times from 0 to x, and by using the given initial conditions $u^{(k)}(0) = b_k$, $0 \leq k \leq (n-1)$ we obtain an expression for $u(x)$ given by

$$u(x) = p(x;\alpha), \tag{11}$$

where $p(x;\alpha)$ is the result derived from integrating (10) and by using the given initial conditions. Substituting (11) into the right hand side of (9), integrating and solving the resulting equation lead to a complete determination of α. The exact solution of (7) follows immediately upon substituting the resulting value of α into (11).

To give a clear view of the technique, we illustrate the method by solving the following examples.

Example 1. Solve the Fredholm integro-differential equation

$$u'(x) = 1 - \frac{1}{3}x + x \int_0^1 t\, u(t)\, dt, \quad u(0) = 0, \tag{12}$$

by using the *direct computation method.*

The equation (12) may be written in the form

$$u'(x) = 1 - \frac{1}{3}x + \alpha\, x, \quad u(0) = 0, \tag{13}$$

where the constant α is defined by

$$\alpha = \int_0^1 t\, u(t)\, dt. \tag{14}$$

To determine α, we first need an expression for $u(x)$ to be used in (14). This can be easily done by integrating both sides of (13) from 0 to x and by using the given initial condition we obtain

$$u(x) = x + \left(\frac{\alpha}{2} - \frac{1}{6}\right) x^2. \tag{15}$$

Substituting (15) into (14) and evaluating the integral yield

$$\alpha = \frac{1}{3}, \tag{16}$$

so that the exact solution

$$u(x) = x, \tag{17}$$

follows immediately upon using (16) into (15).

Example 2. Solve the following Fredholm integro-differential equation

$$u'''(x) = \sin x - x - \int_0^{\pi/2} x t\, u'(t) dt, \tag{18}$$

subject to the initial conditions

$$u(0) = 1, \quad u'(0) = 0, \quad u''(0) = -1, \tag{19}$$

by using the direct computation method.

This equation can be written in the form

$$u'''(x) = \sin x - (1 + \alpha)x, \quad u(0) = 1, \quad u'(0) = 0, \quad u''(0) = -1, \tag{20}$$

where

$$\alpha = \int_0^{\pi/2} t\, u'(t) dt. \tag{21}$$

To determine α, we should find an expression for $u'(x)$ in terms of x and α to be used in Eq. (21). This can be achieved by integrating (20) three times from 0 to x and using the initial conditions, hence we find

$$u''(x) = -\cos x - \frac{1+\alpha}{2!}x^2, \tag{22}$$

$$u'(x) = -\sin x - \frac{1+\alpha}{3!}x^3. \tag{23}$$

and

$$u(x) = \cos x - \frac{1+\alpha}{4!}x^4. \tag{24}$$

Substituting (23) into (21) we obtain

$$\alpha = \int_0^{\pi/2}\left(-t\sin t - \frac{1+\alpha}{3!}t^4\right)dt, \tag{25}$$

which gives

$$\alpha = -1. \tag{26}$$

Substituting (26) into (24) gives

$$u(x) = \cos x, \tag{27}$$

the exact solution in closed form.

Exercises 4.2.1

Solve the following Fredholm integro-differential equations by using the *direct computation method*

1. $u'(x) = \frac{1}{6} + \frac{5}{36}x - \int_0^1 xtu(t)\,dt, \quad u(0) = \frac{1}{6}.$

2. $u'(x) = \frac{1}{21}x - \int_0^1 xtu(t)\,dt, \quad u(0) = \frac{1}{6}\,.$

3. $u''(x) = -\sin x + x - \int_0^{\pi/2} xtu(t)\,dt, \quad u(0) = 0, u'(0) = 1\,.$

4. $u''(x) = \frac{9}{4} - \frac{1}{3}x + \int_0^1 (x-t)u(t)\,dt, \quad u(0) = u'(0) = 0\,.$

5. $u'(x) = 2\sec x^2 \tan x - x + \int_0^{\pi/4} xu(t)\,dt, \quad u(0) = 1.$

4.2.2 The Adomian Decomposition Method:

This method in its simplest form, has been extensively introduced in Chapter 2 for handling Fredholm integral equations. In this section we will study how this powerful method can be implemented to determine a series solution to the Fredholm integro-differential equations. As indicated earlier, we may assume a standard form to the Fredholm integro-differential equation given by

$$u^{(n)}(x) = f(x) + \int_0^1 K(x,t)\,u(t)dt, \quad u^{(k)}(0) = b_k,\ 0 \leq k \leq (n-1) \tag{28}$$

where $u^{(n)}(x)$ indicates the nth derivative of $u(x)$ with respect to x and b_k are constants that give the initial conditions. Substituting (6) into (28) yields

$$u^{(n)}(x) = f(x) + g(x)\int_0^1 h(t)\,u(t)dt. \tag{29}$$

We can easily observe that the definite integral in the integro-differential equation (29) involves an integrand that completely depends on the variable t as discussed in the preceding section. In an operator form, the equation (29) can be written as

$$Lu(x) = f(x) + g(x)\int_0^1 h(t)\,u(t)dt, \tag{30}$$

where the differential operator L is given by

$$L = \frac{d^n}{dx^n}. \tag{31}$$

It is clear that L is an invertible operator, therefore the integral operator L^{-1} is an $n - fold$ integration operator and may be considered as definite integrals from 0 to x for each integral. Applying L^{-1} to both sides of (30) yields

$$\begin{aligned} u(x) &= b_0 + b_1 x + \frac{1}{2!}b_2 x^2 + \cdots + \frac{1}{(n-1)!}\,b_{n-1}x^{n-1} + L^{-1}\left(f(x)\right) \\ &+ \left(\int_0^1 h(t)\,u(t)dt\right)L^{-1}\left(g(x)\right). \end{aligned} \tag{32}$$

In other words we integrated (29) n times from 0 to x and we used the initial conditions at every step of integration. It is important to note that the equation obtained in (32) is a standard Fredholm integral equation. This note will be used in the coming section.

In the decomposition method we usually define the solution $u(x)$ of (28) in a series form given by

$$u(x) = \sum_{n=0}^{\infty} u_n(x). \tag{33}$$

Substituting (33) into both sides of (32) we get

$$\begin{aligned}\sum_{n=0}^{\infty} u_n(x) &= \sum_{k=0}^{n-1} \frac{1}{k!} b_k x^k + L^{-1}(f(x)) \\ &+ \left(\int_0^1 h(t)\,(\sum_{n=0}^{\infty} u_n(t))\,dt\right) L^{-1}(g(x)).\end{aligned} \tag{34}$$

or equivalently

$$\begin{aligned}u_0(x) + u_1(x) + u_2(x) + \cdots &= \sum_{k=0}^{n-1} \frac{1}{k!} b_k x^k + L^{-1}(f(x)) \\ &+ \left(\int_0^1 h(t)\,u_0(t)dt\right) L^{-1}(g(x)) \\ &+ \left(\int_0^1 h(t)\,u_1(t)dt\right) L^{-1}(g(x)) \\ &+ \left(\int_0^1 h(t)\,u_2(t)dt\right) L^{-1}(g(x)) \\ &+ \cdots\end{aligned} \tag{35}$$

The components $u_0(x), u_1(x), u_2(x), u_3(x), \ldots$ of the unknown function $u(x)$ are determined in a recurrent manner, in a similar fashion as discussed before, if we set

$$u_0(x) = \sum_{k=0}^{n-1} \frac{1}{k!} b_k x^k + L^{-1}(f(x)), \tag{36}$$

$$u_1(x) = \left(\int_0^1 h(t)\,u_0(t)dt\right) L^{-1}(g(x)), \tag{37}$$

$$u_2(x) = \left(\int_0^1 h(t)\, u_1(t) dt\right) L^{-1}\left(g(x)\right), \tag{38}$$

$$u_3(x) = \left(\int_0^1 h(t)\, u_2(t) dt\right) L^{-1}\left(g(x)\right), \tag{39}$$

and so on. The above discussed scheme for the determination of the components $u_0(x)$, $u_1(x)$, $u_2(x)$, $u_3(x), \ldots$ of the solution $u(x)$ of the equation (28) can be written in a recursive relationship by

$$u_0(x) = \sum_{k=0}^{n-1} \frac{1}{k!} b_k x^k + L^{-1}\left(f(x)\right), \tag{40}$$

$$u_{n+1}(x) = \left(\int_0^1 h(t)\, u_n(t) dt\right) L^{-1}\left(g(x)\right), \quad n \geq 0. \tag{41}$$

In view of (40) and (41), the components $u_0(x)$, $u_1(x)$, $u_2(x)$, $u_3(x)$, $u_4(x),\ldots$ of $u(x)$ are immediately determined. With these components established, the solution $u(x)$ of (28) is readily determined in a series form using (33). Consequently, the series obtained for $u(x)$ frequently provides the exact solution as will be illustrated later. However, for some problems, where a closed form is not easy to find, we use the series form obtained to approximate the solution. It can be shown that few terms of the series derived by the decomposition method usually provide the higher accuracy level of the approximation.

The decomposition method avoids massive computational work and difficulties that arise from other methods. The computational work can be minimized, sometimes, by observing the so-called self-cancelling noise terms phenomena.

The Noise Terms Phenomena

The phenomena of the self-cancelling noise terms was introduced by [4] and used effectively by [23]. It was proved by [4] and others that the exact solution of any integral or integro-differential equation, for some cases, may be obtained by considering the first two components u_0 and u_1 only. Instead of evaluating several components, it is useful to examine the first two components u_0 and u_1. The conclusion made by [4] suggests that if we observe the appearance of like terms in both components with opposite signs, then by cancelling these terms, the remaining non-cancelled terms of u_0 may in some cases provide the exact solution. This can be justified through substitution. The self-cancelling terms between the components u_0 and u_1 are called the *noise terms*.

It was formally proved that other terms in other components will vanish in the limit if the noise terms occurred in $u_0(x)$ and $u_1(x)$. However, if the exact solution was not attainable by using this phenomena, then we should continue determining other components of $u(x)$ to get a closed form solution or an approximate solution .

Moreover, it is important to note that, even though this is a remarkable achievement that speeds the convergence of the solution and minimizes the size of calculations work, but unfortunately the self cancelling noise terms do not appear always contrary to what [4] tried to prove.

In the following we discuss some examples which illustrate the above outlined decomposition scheme where we will examine the phenomena of the self cancelling noise terms as well.

Example 1. Solve the following Fredholm integro-differential equation

$$u'(x) = \cos x + \frac{1}{4}x - \frac{1}{4}\int_0^{\pi/2} xt\, u(t)dt, \quad u(0) = 0, \tag{42}$$

by using the decomposition method.

Integrating both sides of the equation (42) from 0 to x gives

$$u(x) - u(0) = \sin x + \frac{1}{8}x^2 - \frac{1}{8}x^2 \int_0^{\pi/2} t\, u(t)dt, \quad u(0) = 0, \tag{43}$$

which gives upon using the initial condition

$$u(x) = \sin x + \frac{1}{8}x^2 - \frac{1}{8}x^2 \int_0^{\pi/2} t\, u(t)dt. \tag{44}$$

Using the decomposition technique we usually decompose the solution into a series form given by

$$u(x) = \sum_{n=0}^{\infty} u_n(x). \tag{45}$$

Substituting (45) into both sides of (44) yields

$$\sum_{n=0}^{\infty} u_n(x) = \sin x + \frac{1}{8}x^2 - \frac{1}{8}x^2 \int_0^{\pi/2} t \left(\sum_{n=0}^{\infty} u_n(t)\right) dt, \tag{46}$$

or equivalently

$$\begin{aligned} u_0(x) + u_1(x) + u_2(x) + \cdots = \sin x \; & + \frac{1}{8}x^2 \\ & - \frac{1}{8}x^2 \left(\int_0^{\pi/2} t\, u_0(t)dt\right) \\ & - \frac{1}{8}x^2 \left(\int_0^{\pi/2} t\, u_1(t)dt\right) \\ & - \frac{1}{8}x^2 \left(\int_0^{\pi/2} t\, u_2(t)dt\right) \\ & + \cdots . \end{aligned} \tag{47}$$

Accordingly, we set

$$u_0(x) = \sin x + \tfrac{1}{8}\, x^2, \tag{48}$$

which gives

$$\begin{aligned} u_1(x) &= -\frac{1}{8}\, x^2 \int_0^{\pi/2} t \left(\sin t + \frac{1}{8}t^2\right) dt, \\ &= -\frac{1}{8}\, x^2 - \frac{\pi^4}{16^3}\, x^2. \end{aligned} \tag{49}$$

Considering the first two components $u_0(x)$ and $u_1(x)$ we observe that the two identical terms $\frac{1}{8}x^2$ appear in these components with opposite signs. Cancelling these terms, and substituting the remaining non-cancelled term in $u_0(x)$ to justify that it satisfies the given equation lead to

$$u(x) = \sin x, \tag{50}$$

the exact solution in closed form.

Example 2. Solve the following Fredholm integro-differential equation

$$u'(x) = \frac{1}{6} - \frac{1}{18}x + \int_0^1 xt\, u(t)dt, \quad u(0) = 0, \tag{51}$$

by using the decomposition method.

Integrating both sides of the equation (51) from 0 to x gives

$$u(x) - u(0) = \frac{1}{6}x - \frac{1}{36}x^2 + \frac{1}{2}x^2\left(\int_0^1 t\,u(t)dt\right), \quad u(0) = 0, \tag{52}$$

which gives upon using the initial condition

$$u(x) = \frac{1}{6}x - \frac{1}{36}x^2 + \frac{1}{2}x^2\left(\int_0^1 t\,u(t)dt\right). \tag{53}$$

In the decomposition method we usually express the solution $u(x)$ into a series form given by

$$u(x) = \sum_{n=0}^{\infty} u_n(x). \tag{54}$$

Substituting (54) into both sides of (53) gives

$$\sum_{n=0}^{\infty} u_n(x) = \frac{1}{6}x - \frac{1}{36}x^2 + \frac{1}{2}x^2\left(\int_0^1 t\left(\sum_{n=0}^{\infty} u_n(t)\right)dt\right), \tag{55}$$

or equivalently

$$\begin{aligned} u_0(x) + u_1(x) + u_2(x) + \cdots = \frac{1}{6}x & -\frac{1}{36}x^2 \\ & +\frac{1}{2}x^2\left(\int_0^1 t\,u_0(t)dt\right) \\ & +\frac{1}{2}x^2\left(\int_0^1 t\,u_1(t)dt\right) \\ & +\frac{1}{2}x^2\left(\int_0^1 t\,u_2(t)dt\right) \\ & +\cdots. \end{aligned} \tag{56}$$

Accordingly, we find

$$u_0(x) = \frac{1}{6}x - \frac{1}{36}x^2, \tag{57}$$

$$\begin{aligned} u_1(x) &= \frac{1}{2}x^2\int_0^1 t\left(\frac{1}{6}t - \frac{1}{36}t^2\right)dt, \\ &= \frac{7}{288}x^2. \end{aligned} \tag{58}$$

It is clear that the noise terms do not appear in the two components $u_0(x)$ and $u_1(x)$ explicitly, unless we keep the resulting fractions obtained from integration of (58) distinct. Accordingly, we continue to evaluate more components to get an insight through the solution, hence we obtain

$$\begin{aligned} u_2(x) &= \frac{1}{2}x^2\int_0^1 t\left(\frac{7}{288}t^2\right)dt \\ &= \frac{7}{8(288)}x^2, \end{aligned} \tag{59}$$

and similarly

$$u_3(x) = \frac{7}{64(288)}x^2. \tag{60}$$

Consequently the exact solution can be obtained upon using (54) where we find

$$u(x) = \frac{1}{6}x - \frac{1}{36}x^2 + \frac{7}{288}x^2\left(1+\frac{1}{8}+\frac{1}{64}+\cdots\right), \tag{61}$$

which gives the exact solution

$$u(x) = \frac{1}{6}x, \tag{62}$$

upon evaluating the sum of the infinite geometric series. We point out that the modified decomposition method speeds the procedure to determine the exact solution in this problem by setting $u_0(x) = \frac{1}{6}x$ and continuing as discussed before.

Example 3. Solve the following Fredholm integro-differential equation

$$u'''(x) = \sin x - x - \int_0^{\pi/2} xt\,u'(t)dt, \quad u(0) = 1,\ u'(0) = 0,\ u''(0) = -1 \tag{63}$$

by using the decomposition method.

We point out here that the first derivative $u'(x)$ of the unknown function $u(x)$ appears under the integral sign in this example. The approach we will follow is the same as used before, and will be illustrated through the solution. Integrating both sides of the equation (63) three times from 0 to x yields

$$u(x) - u(0) - \frac{1}{2!}x^2u''(0) = \cos x + \frac{1}{2!}x^2 - \frac{1}{4!}x^4 - 1 - \frac{1}{4!}x^4 \int_0^{\pi/2} t\,u'(t)dt, \tag{64}$$

which gives upon using the initial conditions in (63)

$$u(x) = \cos x - \frac{1}{4!}x^4 - \frac{1}{4!}x^4 \int_0^{\pi/2} t\,u'(t)dt, \tag{65}$$

We begin by expressing the solution $u(x)$ into a series form by

$$u(x) = \sum_{n=0}^{\infty} u_n(x). \tag{66}$$

Substituting (66) into both sides of (65) yields

$$\sum_{n=0}^{\infty} u_n(x) = \cos x - \frac{1}{4!}x^4 - -\frac{1}{4!}x^4 \int_0^{\pi/2} t\left(\sum_{n=0}^{\infty} u_n'(t)\right) dt, \tag{67}$$

or equivalently

$$\begin{aligned} u_0(x) + u_1(x) + u_2(x) + \cdots = \cos x & -\frac{1}{4!}x^4 \\ & -\frac{1}{4!}x^4\left(\int_0^{\pi/2} t\,u_0'(t)dt\right) \\ & -\frac{1}{4!}x^4\left(\int_0^{\pi/2} t\,u_1'(t)dt\right) \\ & -\frac{1}{4!}x^4\left(\int_0^{\pi/2} t\,u_2'(t)dt\right) \\ & + \cdots \end{aligned} \tag{68}$$

Proceeding as before we set

$$u_0(x) = \cos x - \frac{1}{4!}x^4, \tag{69}$$

$$
\begin{aligned}
u_1(x) &= -\frac{1}{4!}x^4 \int_0^{\pi/2} t\left(-\sin t - \frac{1}{3!}t^3\right) dt \\
&= \frac{1}{4!}x^4 + \frac{\pi^5}{(5!)(3!)(32)}x^4.
\end{aligned} \tag{70}
$$

Considering the first two components $u_0(x)$ and $u_1(x)$ in (69) and (70) we observe that the two identical terms $\frac{1}{4!}x^4$ appear in these components with opposite signs. Cancelling these terms and justifying that the remaining non-cancelled term of $u_0(x)$ satisfies the given equation yield the exact solution

$$u(x) = \cos x. \tag{71}$$

We refer the reader to the observation made before in Example 2 that by setting $u_0(x) = \cos x$ and applying the modified decomposition method the solution is easily obtained.

Exercises 4.2.2

Solve the following Fredholm integro-differential equations by using the *decomposition method*

1. $u'(x) = \sinh x + \frac{1}{8}(1 - e^{-1})x - \frac{1}{8}\int_0^1 xtu(t)\,dt, \quad u(0) = 1.$

2. $u'(x) = 1 - \frac{1}{3}x + \int_0^1 xtu(t)\,dt, \quad u(0) = 0.$

3. $u'(x) = xe^x + e^x - x + \int_0^1 xu(t)\,dt, \quad u(0) = 0.$

4. $u'(x) = x\cos x + \sin x - x + \int_0^{\pi/2} xu(t)\,dt, \quad u(0) = 0.$

5. $u''(x) = -\sin x + x - \int_0^{\pi/2} xtu(t)\,dt, \quad u(0) = 0, u'(0) = 1\,.$

6. $u'''(x) = 6+x-\int_0^1 xu''(t)\,dt, \quad u(0) = -1, u'(0) = 1, u''(0) = -2\,.$

7. $u'''(x) = -\cos x + x + \int_0^{\pi/2} xu''(t)\,dt,$

   $u(0) = 0, u'(0) = 1, u''(0) = 0.$

4.2.3 Converting to Fredholm Integral Equations:

In this section we will discuss a technique that will reduce Fredholm integro-differential equation to an equivalent Fredholm integral equation. This can be easily done by integrating both sides of the integro-differential equation as many times as the order of the derivative involved in the equation from 0 to x for every time we integrate, and by using the given initial conditions.

It is important to note that this technique is applicable only if the Fredholm integro-differential equation involves the unknown function $u(x)$ only, and not any of its derivatives, under the integral sign.

Having established the transformation to a standard Fredholm equation, we may proceed using any of the alternative methods that were discussed before in Chapter 2, namely the decomposition method, the direct computation method, the successive approximations method or the method of successive substitutions.

To give a clear overview of this method we discuss the following illustrative examples.

Example 1. Solve the following Fredholm integro-differential equation

$$u'(x) = 1 - \frac{1}{3}x + x\int_0^1 t\,u(t)dt, \quad u(0) = 0, \tag{72}$$

by converting it to a standard Fredholm integral equation.

Integrating both sides from 0 to x and using the initial condition we obtain

$$u(x) = x - \frac{1}{3!}x^2 + \frac{1}{2!}x^2\left(\int_0^1 t\,u(t)dt\right). \tag{73}$$

It can be easily seen that (73) is a Fredholm integral equation; therefore we can select any method of the alternative methods that were introduced before. For variety we select the successive approximations method to solve this equation. Hence we set a zeroth approximation by

$$u_0(x) = x, \tag{74}$$

and using this choice in (73) yields the first approximation

$$u_1(x) = x - \frac{1}{3!}x^2 + \frac{1}{2!}x^2\left(\int_0^1 t^2 dt\right), \tag{75}$$

which gives

$$u_1(x) = x. \tag{76}$$

It is obvious that if we continue in the same manner, we then obtain

$$u_n(x) = x. \tag{77}$$

Accordingly,

$$\begin{aligned} u(x) &= \lim_{n\to\infty} u_n(x) \\ &= \lim_{n\to\infty} x \\ &= x. \end{aligned} \tag{78}$$

Example 2. Solve the following Fredholm integro-differential equation

$$u''(x) = e^x - x + x\int_0^1 t\,u(t)dt, \quad u(0) = 1, \quad u'(0) = 1, \tag{79}$$

by reducing it to a Fredholm integral equation.

Integrating both sides of (79) twice from 0 to x and using the initial conditions we obtain

$$u(x) = e^x - \frac{1}{3!}x^3 + \frac{1}{3!}x^3\int_0^1 tu(t)\,dt, \tag{80}$$

a typical Fredholm integral equation. As indicated earlier we can select any method that will determine the solution; therefore we will use the direct computation method for this example. Therefore we can express (80) in the form

$$u(x) = e^x - \frac{1}{3!}x^3 + \alpha\frac{1}{3!}x^3, \tag{81}$$

where the constant α is defined by the definite integral

$$\alpha = \int_0^1 tu(t)\,dt. \tag{82}$$

Substituting (81) into (82) we obtain

$$\alpha = \int_0^1 t\left(e^t - \frac{1}{3!}t^3 + \alpha\frac{1}{3!}t^3\right)\,dt, \tag{83}$$

an easy integral to evaluate, from which we obtain

$$\alpha = 1. \tag{84}$$

Inserting the value of α obtained in (84) into (81) yields the exact solution given by

$$u(x) = e^x. \tag{85}$$

In closing this section, the main ideas we applied are the direct computation method and the decomposition method, where the noise terms phenomena was introduced. The direct computation method provides the solution in a closed form, but the decomposition method provides the solution in a rapid convergent series as discussed above.

Besides, the integro-differential equation can be easily reduced to standard Fredholm integral equation if the unknown function $u(x)$, and not any derivative of $u(x)$, appears under the integral sign. Converting to Fredholm equation, then we can use any of the alternative methods discussed before in Chapter 2.

Exercises 4.2.3

Solve the following Fredholm integro-differential equations by *converting it to Fredholm integral equations*

1. $u'(x) = -x\sin x + \cos x + (1-\pi/2)x + \int_0^{\pi/2} xu(t)\,dt, \quad u(0) = 0.$

2. $u''(x) = -e^x + \frac{1}{2}x + \int_0^1 xtu(t)\,dt, \quad u(0) = 0, u'(0) = -1.$

3. $u''(x) = -\sin x + \cos x + (2-\pi/2)x - \int_0^{\pi/2} xtu(t)\,dt,$

   $u(0) = -1, u'(0) = 1.$

4. $u'(x) = \frac{7}{6} - 11x - \int_0^1 (x-t)u(t)\,dt, \quad u(0) = 0\ .$

5. $u'(x) = \frac{1}{4} + \cos(2x) - \int_0^{\pi/4} xu(t)\,dt, \quad u(0) = 0\ .$

4.3 Volterra Integro-Differential Equations

In this section we will present the reliable methods that will be used to handle Volterra integro-differential equations. We will focus our study on equations that involve separable kernels of the form

$$K(x,t) = \sum_{k=1}^{n} g_k(x)\, h_k(t). \tag{86}$$

Without loss of generality, we will consider the cases where the kernel $K(x,t)$ consists of one product of the functions $g(x)$ and $h(t)$ given by

$$K(x,t) = g(x)\, h(t), \tag{87}$$

where other cases can be generalized in the same manner. The nonseparable kernel can be reduced to separable kernel by using the Taylor expansion for the kernel involved. The methods to be introduced are identical, with some exceptions, to the methods discussed in Chapter 3. Our approach will be mainly based on how we can extend the methods used in Chapter 3 to handle this type of equations. For this reason we first start with the most practical method.

4.3.1 The Series Solution Method:

This method has been extensively introduced in Chapter 3. Without loss of generality, we may consider a standard form to the Volterra integro-differential equation given by

$$u^{(n)}(x) = f(x) + \int_0^x K(x,t)\, u(t)dt, \quad u^{(k)}(0) = b_k,\ 0 \le k \le (n-1), \tag{88}$$

where $u^{(n)}(x)$ indicates the nth derivative of $u(x)$ with respect to x, and b_k are constants that define the initial conditions. Substituting (87) into (88) yields

$$u^{(n)}(x) = f(x) + g(x)\int_0^x h(t)\, u(t)dt, \quad u^{(k)}(0) = b_k,\ 0 \le k \le (n-1) \tag{89}$$

We should follow a parallel manner to the approach of the *series solution method* that usually used in solving ordinary differential equations around an ordinary point. To achieve this goal, we first assume that the solution $u(x)$ of (89) is an analytic function and hence can be represented by a series expansion given by

$$u(x) = \sum_{k=0}^{\infty} a_k x^k, \tag{90}$$

where the coefficients a_k are constants that will be determined. It is to be noted that the first few coefficients a_k can be determined by using the initial conditions so that

$$a_0 = u(0), \tag{91}$$

$$a_1 = u'(0), \tag{92}$$

$$a_2 = \frac{1}{2!} u''(0), \tag{93}$$

and so on depending on the number of the initial conditions ,whereas the remaining coefficients a_k will be determined from applying the technique as will be discussed later . Substituting (90) into both sides of (89) yields

$$\left(\sum_{k=0}^{\infty} a_k x^k\right)^{(n)} = f(x) + g(x) \int_0^x h(t) \left(\sum_{k=0}^{\infty} a_k t^k\right) dt. \tag{94}$$

In view of (94), Eq. (89) will be reduced to calculable integrals in the right hand side of Eq. (94) that can be easily evaluated where we have to integrate terms of the form t^n, $n \geq 0$ only.

The next step is to write the Taylor expansion for $f(x)$, evaluate the resulting traditional integrals in (94), and then equating the coefficients of like powers of x in both sides of the equation. This will lead to a complete determination of the coefficients $a_0, a_1, a_2, \ldots$ of the series in (90).

Consequently, substituting the obtained coefficients a_k, $k \geq 0$ in (90) produces the solution in a series form. This may give a solution in a closed form, if the expansion obtained is a Taylor expansion to a well known elementary function, or we may use a series form solution if a closed form is not attainable.

To give a clear overview of the technique and how it should be implemented for Volterra integro-differential equations, the series solution method will be illustrated by discussing the following examples.

Example 1. Solve the following Volterra integro-differential equation

$$u''(x) = x\cosh x - \int_0^x t\,u(t)\,dt, \quad u(0) = 0, u'(0) = 1, \tag{95}$$

by using the series solution method.

Substituting $u(x)$ by the series

$$u(x) = \sum_{n=0}^{\infty} a_n x^n, \tag{96}$$

into both sides of the equation (95) and using the Taylor expansion of $\cosh x$ we obtain

$$\sum_{n=2}^{\infty} n(n-1)a_n x^{n-2} = x\left(\sum_{k=0}^{\infty} \frac{x^{2k}}{(2k)!}\right) - \int_0^x t\left(\sum_{n=0}^{\infty} a_n t^n\right) dt. \tag{97}$$

Using the initial conditions yields

$$a_0 = 0, \tag{98}$$

$$a_1 = 1. \tag{99}$$

Evaluating the integrals that involve terms of the form t^n, $n \geq 0$, and using few terms from both sides yield

$$\begin{aligned} 2a_2 + 6a_3 x + 12a_4 x^2 + 20a_5 x^3 + \cdots &= x\left(1 + \frac{1}{2!}x^2 + \frac{1}{4!}x^4 + \cdots\right) \\ &\quad - \left(\frac{1}{3}x^3 + \frac{1}{4}a_2 x^4 + \cdots\right). \end{aligned} \tag{100}$$

Equating the coefficients of like powers of x in both sides we find

$$a_2 = 0, \tag{101}$$

$$a_3 = \frac{1}{3!}, \tag{102}$$

$$a_4 = 0, \tag{103}$$

and generally

$$a_{2n} = 0, \quad for \quad n \geq 0, \tag{104}$$

and

$$a_{2n+1} = \frac{1}{(2n+1)!}, \quad for \quad n \geq 0. \tag{105}$$

Using (96) we find the solution $u(x)$ in a series form

$$u(x) = x + \frac{1}{3!}x^3 + \frac{1}{5!}x^5 + \frac{1}{7!}x^7 + \ldots, \tag{106}$$

and in a closed form

$$u(x) = \sinh x, \tag{107}$$

is the exact solution of Eq. (95).

Example 2. As a second example we use the series solution method to solve

$$u''(x) = \cosh x + \frac{1}{4} - \frac{1}{4}\cosh 2x + \int_0^x \sinh t u(t)dt, \quad u(0) = 1, \, u'(0) = 0. \tag{108}$$

Using (90) – (92) and considering the first few terms of the expansion of $u(x)$, we obtain

$$u(x) = 1 + a_2x^2 + a_3x^3 + a_4x^4 + a_5x^5 + \cdots. \tag{109}$$

Substituting (109) into both sides of (108) yields

$$\begin{aligned} 2a_2 \; + \; & 6a_3x + 12a_4x^2 + 20a_5x^3 + \cdots = \left(1 + \frac{x^2}{2!} + \frac{x^4}{4!} + \cdots\right) + \frac{1}{4} \\ & - \frac{1}{4}\left(1 + \frac{(2x)^2}{2!} + \frac{(2x)^4}{4!} + \cdots\right) \\ & + \int_0^x \left(t + \frac{t^3}{3!} + \frac{t^5}{5!} + \cdots\right)\left(1 + a_2t^2 + a_3t^3 + \cdots\right) dt. \end{aligned} \tag{110}$$

Integrating the right hand side and equating the coefficients of like powers of x we find

$$a_0 = 1, \tag{111}$$

$$a_1 = 0, \tag{112}$$

$$a_2 = \frac{1}{2!}, \tag{113}$$

$$a_3 = 0, \tag{114}$$

$$a_4 = \frac{1}{4!}, \tag{115}$$

$$a_5 = 0, \tag{116}$$

and so on, where the constants a_0 and a_1 are defined by using the initial conditions. Consequently the solution in a series form is given by

$$u(x) = 1 + \frac{x^2}{2!} + \frac{x^4}{4!} + \frac{x^6}{6!} + \cdots, \tag{117}$$

which gives

$$u(x) = \cosh x, \tag{118}$$

the exact solution in a closed form.

Exercises 4.3.1

Solve the following Volterra integro-differential equations by using *the series solution method*

1. $u'(x) = 1 - 2x\sin x + \int_0^x u(t)\,dt, \quad u(0) = 0.$

2. $u'(x) = -1 + \frac{1}{2}x^2 - xe^x - \int_0^x tu(t)\,dt, \quad u(0) = 0\ .$

3. $u''(x) = 1 - x(\cos x + \sin x) - \int_0^x tu(t)\,dt, \quad u(0) = -1, u'(0) = 1.$

4. $u''(x) = -8 - \frac{1}{3}(x^3 - x^4) + \int_0^x (x-t)u(t)\,dt, \quad u(0) = 0, u'(0) = 2.$

$$5.\ u''(x) = \frac{1}{2}x^2 - x\cosh x - \int_0^x tu(t)\,dt, \quad u(0) = 1, u'(0) = -1.$$

4.3.2 The Decomposition Method:

The decomposition method and the modified decomposition method were discussed in details in Chapter 3. In this section we will introduce how this successful method can be implemented to determine a series solution to the Volterra integro-differential equations. As will be seen later, the method is reliable and effective.

Without loss of generality, we may assume a standard form to the Volterra integro-differential equation defined by the standard form

$$u^{(n)}(x) = f(x) + \int_0^x K(x,t)\,u(t)dt, \quad u^{(k)}(0) = b_k,\ 0 \leq k \leq (n-1) \tag{119}$$

where $u^{(n)}(x)$ indicates the nth derivative of $u(x)$ with respect to x and b_k are constants that define the initial conditions. It is natural to seek an expression for $u(x)$ that will be derived from (119). This can be done by integrating both sides of (119) from 0 to x as many times as the order of the derivative involved. Consequently, we obtain

$$u(x) = \sum_{k=0}^{n-1} \frac{1}{k!} b_k x^k + L^{-1}(f(x)) + L^{-1}\left(\int_0^x K(x,t)\,u(t)dt\right), \tag{120}$$

where $\sum_{k=0}^{n-1} \frac{1}{k!} b_k x^k$ is obtained by using the initial conditions, and L^{-1} is an n-fold integration operator. Now we apply the decomposition method by defining the solution $u(x)$ of (120) in a decomposition series given by

$$u(x) = \sum_{n=0}^{\infty} u_n(x). \tag{121}$$

Substituting (121) into both sides of (120) we get

$$\sum_{n=0}^{\infty} u_n(x) = \sum_{k=0}^{n-1} \frac{1}{k!} b_k x^k + L^{-1}(f(x)) + L^{-1}\left(\int_0^x K(x,t)\left(\sum_{n=0}^{\infty} u_n(t)\right) dt\right) \tag{122}$$

or equivalently

$$\begin{aligned} u_0(x) + u_1(x) + u_2(x) + \cdots &= \sum_{k=0}^{n-1} \frac{1}{k!} b_k x^k + L^{-1}(f(x)) \\ &+L^{-1}\left(\int_0^x K(x,t)\, u_0(t) dt\right) \\ &+L^{-1}\left(\int_0^x K(x,t)\, u_1(t) dt\right) \\ &+L^{-1}\left(\int_0^x K(x,t)\, u_2(t) dt\right) \\ &+\cdots. \end{aligned} \tag{123}$$

The components $u_0(x), u_1(x), u_2(x), u_3(x), \ldots$ of the unknown function $u(x)$ are determined in a recursive manner, in a similar way as discussed before, if we set

$$u_0(x) = \sum_{k=0}^{n-1} \frac{1}{k!} a_k x^k + L^{-1}(f(x)), \tag{124}$$

$$u_1(x) = L^{-1}\left(\int_0^x K(x,t)\, u_0(t) dt\right), \tag{125}$$

$$u_2(x) = L^{-1}\left(\int_0^x K(x,t)\, u_1(t) dt\right), \tag{126}$$

$$u_3(x) = L^{-1}\left(\int_0^x K(x,t)\, u_2(t) dt\right), \tag{127}$$

and so on. The decomposition method discussed above for the determination of the components $u_0(x), u_1(x), u_2(x), u_3(x), \ldots$ of the solution $u(x)$ of the equation (119) can be written in a recursive manner by

$$u_0(x) = \sum_{k=0}^{n-1} \frac{1}{k!} a_k x^k + L^{-1}(f(x)), \tag{128}$$

$$u_{n+1}(x) = L^{-1}\left(\int_0^x K(x,t)\, u_n(t) dt\right), \quad n \geq 0. \tag{129}$$

In view of (128) and (129), the components $u_0(x)$, $u_1(x)$, $u_2(x)$,... are immediately determined. With the components determined, the solution $u(x)$ of (119) is then obtained in a series form using (121). Consequently, the series obtained for $u(x)$ mostly provides the exact solution in a closed form as will be illustrated later. However, for concrete problems, where (121) cannot be evaluated, a truncated series $\sum_{n=0}^{k} u_n(x)$ is usually used to approximate the solution $u(x)$.

It is convenient to point out that the phenomena of the self-cancelling noise terms that was introduced before may be applied here if the noise terms appear in $u_0(x)$ and $u_1(x)$. The following examples will explain how we can use the decomposition technique.

Example 1. Solve the following Volterra integro-differential equation

$$u''(x) = x + \int_0^x (x-t)\,u(t)dt, \quad u(0) = 0, u'(0) = 1, \tag{130}$$

by using the decomposition method.

Applying the *two − fold* integration operator L^{-1}

$$L^{-1}(.) = \int_0^x \int_0^x (.)dx\,dx, \tag{131}$$

to both sides of (130), i.e. integrating both sides of (130) twice from 0 to x, and using the given initial conditions yield

$$u(x) = x + \frac{1}{3!}x^3 + L^{-1}\left(\int_0^x (x-t)\,u(t)dt\right). \tag{132}$$

Following the decomposition scheme (128) and (129) we find

$$u_0(x) = x + \frac{1}{3!}x^3, \tag{133}$$

$$\begin{aligned} u_1(x) &= L^{-1}\left(\int_0^x (x-t)\,u_0(t)dt\right), \\ &= \frac{1}{5!}x^5 + \frac{1}{7!}x^7, \end{aligned} \tag{134}$$

$$\begin{aligned} u_2(x) &= L^{-1}\left(\int_0^x (x-t)\,u_1(t)dt\right), \\ &= \frac{1}{9!}x^9 + \frac{1}{11!}x^{11}. \end{aligned} \tag{135}$$

Combining the equations (133) – (135) yields the solution $u(x)$ in a series form given by

$$u(x) = x + \frac{1}{3!}x^3 + \frac{1}{5!}x^5 + \frac{1}{7!}x^7 + \frac{1}{9!}x^9 + \frac{1}{11!}x^{11} + \cdots, \qquad (136)$$

and this leads to

$$u(x) = \sinh x, \qquad (137)$$

the exact solution in a closed form.

Example 2. Solve the following Volterra integro-differential equation

$$u''(x) = 1 + \int_0^x (x-t)\, u(t)dt, \quad u(0) = 1, u'(0) = 0, \qquad (138)$$

by using the decomposition method. Integrating both sides of (138) twice from 0 to x and using the given initial conditions yield

$$u(x) = 1 + \frac{1}{2!}x^2 + L^{-1}\left(\int_0^x (x-t)\, u(t)dt\right). \qquad (139)$$

where L^{-1} is a two-fold integration operator given above by (131). Following the decomposition method we obtain

$$u_0(x) = 1 + \frac{1}{2!}x^2, \qquad (140)$$

$$u_1(x) = L^{-1}\left(\int_0^x (x-t)\, u_0(t)dt\right) = \frac{1}{4!}x^4 + \frac{1}{6!}x^6, \qquad (141)$$

$$u_2(x) = L^{-1}\left(\int_0^x (x-t)\, u_1(t)dt\right) = \frac{1}{8!}x^8 + \frac{1}{10!}x^{10}. \qquad (142)$$

Combining the results (140) – (142) yields the solution $u(x)$ in a series form given by

$$u(x) = 1 + \frac{1}{2!}x^2 + \frac{1}{4!}x^4 + \frac{1}{6!}x^6 + \frac{1}{8!}x^8 + \frac{1}{10!}x^{10} + \cdots, \qquad (143)$$

and this gives

$$u(x) = \cosh x, \tag{144}$$

the exact solution in a closed form.

Example 3. Solve the following Volterra integro-differential equation

$$u'''(x) = -1 + \int_0^x u(t)dt, \quad u(0) = u'(0) = 1, u''(0) = -1, \tag{145}$$

by using the decomposition method.

We note here that the Volterra integro-differential equation involves the third order differential operator $u'''(x)$, therefore integrating both sides of (145) three times from 0 to x and using the initial conditions we obtain

$$u(x) = 1 + x - \frac{1}{2!}x^2 - \frac{1}{3!}x^3 + L^{-1}\left(\int_0^x u(t)dt\right). \tag{146}$$

Following the decomposition scheme we find

$$u_0(x) = 1 + x - \frac{1}{2!}x^2 - \frac{1}{3!}x^3, \tag{147}$$

which gives

$$\begin{aligned} u_1(x) &= L^{-1}\left(\int_0^x u_0(t)dt\right), \\ &= \frac{1}{4!}x^4 + \frac{1}{5!}x^5 - \frac{1}{6!}x^6 - \frac{1}{7!}x^7. \end{aligned} \tag{148}$$

Consequently, the solution $u(x)$ given in a series form

$$u(x) = 1 + x - \frac{1}{2!}x^2 - \frac{1}{3!}x^3 + \frac{1}{4!}x^4 + \frac{1}{5!}x^5 - \frac{1}{6!}x^6 - \frac{1}{7!}x^7 + \cdots \tag{149}$$

We can easily observe that the series solution obtained in (149) will not easily give the closed form solution; however, rewriting (149) by

$$u(x) = \left(1 - \frac{1}{2!}x^2 + \frac{1}{4!}x^4 + \cdots\right) + \left(x - \frac{1}{3!}x^3 + \frac{1}{5!}x^5 + \cdots\right), \tag{150}$$

gives

$$u(x) = \cos x + \sin x, \tag{151}$$

the closed form solution.

Exercises 4.3.2

Solve the following Volterra integro-differential equations by using *the Adomian decomposition method*

1. $u''(x) = 1 + x - \frac{1}{3!}x^3 + \int_0^x (x-t)u(t)\,dt, \quad u(0) = 1, u'(0) = 2$.

2. $u''(x) = -1 - \frac{1}{2!}x^2 + \int_0^x (x-t)u(t)\,dt, \quad u(0) = 2, u'(0) = 0.$

3. $u'(x) = 2 + \int_0^x u(t)\,dt, \quad u(0) = 2.$

4. $u'(x) = 1 - \int_0^x u(t)\,dt, \quad u(0) = 1.$

5. $u^{(iv)}(x) = -x + \frac{1}{2!}x^2 - \int_0^x (x-t)u(t)\,dt,$

$$u(0) = 1, u'(0) = -1, u''(0) = 0, u'''(0) = 1.$$

4.3.3 Converting to Volterra Integral Equation:

We can easily convert the Volterra integro-differential equation to an equivalent Volterra integral equation, provided that the kernel is a *difference kernel* defined by the form $K(x,t) = K(x-t)$. This can be easily done by integrating both sides of the equation and using the initial conditions. To perform the conversion to a regular Volterra integral equation we should used the formula (55) of Chapter 1 that converts multiple integral to a single integral. The reader is advised to review that formula for further reference. The following two specific formulas

$$\int_0^x \int_0^x u(t)\,dt\,dt = \int_0^x (x-t)u(t)\,dt, \tag{152}$$

and

$$\int_0^x \int_0^x \int_0^x u(t)\,dt\,dt\,dt = \frac{1}{2!}\int_0^x (x-t)^2 u(t)\,dt. \tag{153}$$

given by (56) and (57) in Chapter 1 are usually used to transform double integrals and triple integrals respectively to a single integral. Having established the transformation to a standard Volterra integral equation, we may proceed using any of the alternative methods that were discussed before in Chapter 3.

To give a clear overview of this method we discuss the following examples.

Example 1. Solve the following Volterra integro-differential equation

$$u'(x) = 2 - \frac{1}{4}x^2 + \frac{1}{4}\int_0^x u(t)dt, \quad u(0) = 0, \tag{154}$$

by converting to a standard Volterra integral equation.

Integrating both sides from 0 to x and using the initial condition we obtain

$$u(x) = 2x - \frac{1}{12}x^3 + \frac{1}{4}\int_0^x \int_0^x u(t)\,dt\,dt, \tag{155}$$

which gives

$$u(x) = 2x - \frac{1}{12}x^3 + \frac{1}{4}\int_0^x (x-t)\,u(t)\,dt, \tag{156}$$

upon using the formula (152). It is clearly seen that (156) is a standard Volterra integral equation that will be solved by using the decomposition method. Following that technique we set

$$u_0(x) = 2x - \frac{1}{12}x^3, \tag{157}$$

which gives

$$u_1(x) = \frac{1}{4}\int_0^x (x-t)\left(2t - \frac{1}{12}t^3\right)dt, \tag{158}$$

so that

$$u_1(x) = \frac{1}{12}x^3 - \frac{1}{240}x^5. \tag{159}$$

We can easily observe that the term $\frac{1}{12}x^3$ appears with opposite signs in the components $u_0(x)$ and $u_1(x)$, and by cancelling this noise term from $u_0(x)$ and justifying that

$$u(x) = 2x, \tag{160}$$

is the exact solution of (156).

Example 2. Solve the following Volterra integro-differential equation

$$u''(x) = -1 + \int_0^x (x-t)u(t)dt, \quad u(0) = 1, u'(0) = 0, \tag{161}$$

by converting to a standard Volterra integral equation.

Integrating both sides twice from 0 to x and using the initial condition we obtain

$$\begin{aligned} u'(x) &= -x + \int_0^x \int_0^x (x-t)\,u(t)\,dt\,dt, \\ &= -x + \frac{1}{2!}\int_0^x (x-t)^2 u(t)dt, \end{aligned} \tag{162}$$

by using formula (152) ,and

$$\begin{aligned} u(x) &= 1 - \frac{1}{2!}x^2 + \int_0^x \int_0^x \int_0^x (x-t)\,u(t)\,dt\,dt\,dt, \\ &= 1 - \frac{1}{2!}x^2 + \frac{1}{3!}\int_0^x (x-t)^3 u(t)dt, \end{aligned} \tag{163}$$

by using formula (153). The last equation is a standard Volterra integral equation that will be solved by using the modified decomposition method. To determine $u(x)$ we set

$$u_0(x) = 1, \tag{164}$$

which gives

$$u_1(x) = -\frac{1}{2!}x^2 + \frac{1}{4!}x^4. \tag{165}$$

We can easily observe that the noise terms did not appear between the components $u_0(x)$ and $u_1(x)$, therefore we continue to find more terms to study the solution more closely. Consequently, we find

$$u_2(x) = -\frac{1}{6!}x^6 + \frac{1}{8!}x^8. \tag{166}$$

Combining the results for the components $u_0(x)$, $u_1(x)$ and $u_2(x)$ we obtain the series expression for the solution given by

$$u(x) = 1 - \frac{1}{2!}x^2 + \frac{1}{4!}x^4 - \frac{1}{6!}x^6 + \frac{1}{8!}x^8 + \cdots, \tag{167}$$

so that

$$u(x) = \cos x, \tag{168}$$

is the closed form of the exact solution

Exercises 4.3.3

Solve the following Volterra integro-differential equations by converting the problem to *Volterra integral equation*

1. $u''(x) = 1 + \int_0^x (x-t)u(t)\,dt, \quad u(0) = 1, u'(0) = 0.$

2. $u'(x) = 1 - \int_0^x u(t)\,dt, \quad u(0) = 0\ .$

3. $u''(x) = x + \int_0^x (x-t)u(t)\,dt, \quad u(0) = 0, u'(0) = 1\ .$

4. $u'(x) = 2 - \frac{1}{2!}x^2 + \int_0^x u(t)\,dt, \quad u(0) = 1.$

5. $u'(x) = 1 - \int_0^x u(t)\,dt, \quad u(0) = 1$

6. $u''(x) = 1 + x + \int_0^x (x-t)u(t)\,dt, \quad u(0) = u'(0) = 1.$

4.3.4 Converting to Initial Value Problems:

In this section we will study how to reduce the Volterra integro-differential equation to an equivalent initial value problem, focusing our discussion on the case where the kernel is a difference kernel of the form $K(x,t) = K(x-t)$. This can be achieved easily by differentiating both sides of the integro-differential equation as many times as needed to remove the integral sign. In differentiating the integral involved we shall use the Leibnitz rule to achieve our goal. The Leibnitz rule has been introduced in Section 1.4. It is important to note that we should define the initial conditions at every step of differentiation. A similar technique was discussed and examined in Chapters 1 and 2. The reader is advised to review the related material for further use.

Having converted the Volterra integro-differential equation to an initial value problem, the various methods that are used in any ordinary differential equation course can be used to determine the solution. The idea is easy to use but requires more calculations if compared with the integral equations techniques.

To give a clear overview of this method we discuss the following illustrative examples.

Example 1. Solve the following Volterra integro-differential equation

$$u'(x) = 1 + \int_0^x u(t)dt, \quad u(0) = 0, \tag{169}$$

by converting it to an initial value problem.

Differentiating both sides of (169) with respect to x and using the Leibnitz rule to differentiate the integral at the right hand side we obtain

$$u''(x) = u(x), \tag{170}$$

with initial conditions given by

$$u(0) = 0, \qquad u'(0) = 1, \tag{171}$$

where the last initial condition was obtained by substituting $x = 0$ in both sides of (169). The characteristic equation of (170) is

$$r^2 - 1 = 0, \tag{172}$$

which gives the roots

$$r = \pm 1, \tag{173}$$

so that the general solution is given by

$$u(x) = A\cosh x + B\sinh x, \tag{174}$$

where A and B are constants to be determined. Using the initial conditions given by (171) to find the constants A and B, we find that

$$u(x) = \sinh x, \tag{175}$$

is the exact solution.

Example 2. Solve the following Volterra integro-differential equation

$$u''(x) = -x + \int_0^x (x-t)u(t)dt, \quad u(0) = 0,\ u'(0) = 1, \tag{176}$$

by converting it to an initial value problem.

Differentiating both sides of (176) and using Leibnitz rule we find

$$u'''(x) = -1 + \int_0^x u(t)\,dt, \tag{177}$$

and by differentiating again to reduce the equation to a pure differential equation we obtain

$$u^{(iv)}(x) = u(x). \tag{178}$$

Combining the given initial conditions in (176) with the other initial conditions, obtained by substituting $x = 0$ in (176) and (177), we write

$$u(0) = 0,\ u'(0) = 1,\ u''(0) = 0,\ u'''(0) = -1. \tag{179}$$

The characteristic equation of (178) is

$$r^4 - 1 = 0, \tag{180}$$

which gives the roots

$$r = \pm 1, \pm \imath. \tag{181}$$

so that the general solution is given by

$$u(x) = A\cosh x + B\sinh x + C\cos x + D\sin x, \tag{182}$$

where A , B ,C and D are constants to be determined. Using the initial conditions (179) to determine the numerical values for the constants A, B, C and D, we find that

$$u(x) = \sin x, \tag{183}$$

is the solution of the problem under discussion.

Example 3. Solve the following Volterra integro-differential equation

$$u'(x) = 2 - \frac{1}{4}x^2 + \frac{1}{4}\int_0^x u(t)dt \quad u(0) = 0, \tag{184}$$

by reducing the equation to an initial value problem.

Differentiating both sides of (184) with respect to x and using Leibnitz rule to differentiate the integral at the right hand side we obtain

$$u''(x) = -\frac{1}{2}x + \frac{1}{4}u(x), \tag{185}$$

or equivalently

$$u''(x) - \frac{1}{4}u(x) = -\frac{1}{2}x, \tag{186}$$

with initial conditions given by

$$u(0) = 0, \quad u'(0) = 2, \tag{187}$$

where the second initial condition was obtained by substituting $x = 0$ in both sides of (184). The characteristic equation for the corresponding homogeneous equation of (186) is

$$r^2 - \frac{1}{4} = 0, \tag{188}$$

which gives the roots

$$r = \pm\frac{1}{2}, \tag{189}$$

so that the complementary solution is given by

$$u_c(x) = A\cosh(\frac{x}{2}) + B\sinh(\frac{x}{2}), \tag{190}$$

where A and B are constants to be determined. A particular solution $u_p(x)$ can be obtained by assuming that

$$u_p(x) = C + Dx. \tag{191}$$

Substituting (191) into (186) yields

$$C = 0, \quad D = 2. \tag{192}$$

Combining (190) – (192) yields

$$u(x) = A\cosh(\frac{x}{2}) + B\sinh(\frac{x}{2}) + 2x, \tag{193}$$

which gives

$$u(x) = 2x, \tag{194}$$

upon using the initial conditions (187) .

Students are advised to make more revision of solving homogeneous and nonhomogeneous ordinary differential equations with constant coefficients.

Exercises 4.3.4

Solve the following Volterra integro-differential equations by converting the problem to *an initial value problem*

1. $u'(x) = e^x - \int_0^x u(t)\,dt, \quad u(0) = 1.$

2. $u'(x) = 1 - \int_0^x u(t)\,dt, \quad u(0) = 0.$

3. $u''(x) = -x - \frac{1}{2!}x^2 + \int_0^x (x-t)u(t)\,dt, \quad u(0) = 1, u'(0) = 1.$

4. $u''(x) = 1 - \frac{1}{2!}x^2 + \int_0^x (x-t)u(t)\,dt, \quad u(0) = 2, u'(0) = 0.$

5. $u''(x) = -\frac{1}{2!}x^2 - \frac{2}{3}x^3 + \int_0^x (x-t)u(t)\,dt, \quad u(0) = 1, u'(0) = 4.$

6. $u''(x) = -x - \frac{1}{8}x^2 + \int_0^x (x-t)u(t)\,dt, \quad u(0) = \frac{1}{4}, u'(0) = 1.$

7. $u'(x) = 1 + \sin x + \int_0^x u(t)\,dt, \quad u(0) = -1.$

Chapter 5

Singular Integral Equations

5.1 Definitions

An integral equation is called a *singular* integral equation if one or both limits of integration become *infinite*, or if the kernel K(x,t) of the equation becomes *infinite* at one or more points in the interval of integration. In other words, the integral equation of the first kind

$$f(x) = \lambda \int_{\alpha(x)}^{\beta(x)} K(x,t)\, u(t)dt. \tag{1}$$

or the integral equation of the second kind

$$u(x) = f(x) + \lambda \int_{\alpha(x)}^{\beta(x)} K(x,t)\, u(t)dt, \tag{2}$$

is called *singular* if the lower limit $\alpha(x)$, the upper limit $\beta(x)$ or both limits of integration are *infinite*. Moreover, the equation (1) or (2) is also called a *singular* integral equation if the kernel $K(x,t)$ becomes *infinite* at one or more points in the domain of integration. Examples of the first type of *singular* integral equations are given by the following examples:

$$u(x) = 1 + e^{-x} - \int_0^{\infty} u(t)dt, \tag{3}$$

$$F(\lambda) = \int_{-\infty}^{\infty} e^{-\imath\lambda x} u(x) dx, \tag{4}$$

$$L[u(x)] = \int_{0}^{\infty} e^{-\lambda x} u(x) dx. \tag{5}$$

The integral equations (4) and (5) are Fourier transform and Laplace transform of the function $u(x)$ respectively. In addition, the equations (4) and (5) are in fact Fredholm integral equations of the first kind with kernels given by $K(x,t) = e^{-\imath\lambda x}$ and $K(x,t) = e^{-\lambda x}$. It is important to note that the Laplace transforms and the Fourier transforms are usually used for solving ordinary and partial differential equations with constant coefficients. However, these transforms will not be used in this text to solve integral equations, but will be used in the derivation of two formulas as will be seen later.

One important point to be noted here is that the singular behavior in (3) – (5) has been attributed to the domain of integration becoming *infinite.*

Examples of the second type of *singular* integral equations are given by the following

$$x^2 = \int_0^x \frac{1}{\sqrt{x-t}} u(t) dt, \tag{6}$$

$$x = \int_0^x \frac{1}{(x-t)^{\alpha}} u(t) dt, \quad 0 < \alpha < 1, \tag{7}$$

$$u(x) = 1 + 2\sqrt{x} - \int_0^x \frac{1}{\sqrt{x-t}} u(t) dt, \tag{8}$$

where the singular behavior in this style of equations has been attributed to the kernel K(x,t) becoming *infinite* as $t \to x$.

It is important to note that integral equations similar to examples (6) and (7) are called Abel's problems and generalized Abel's integral equations respectively. Moreover these styles of singular integral equations are among the earliest integral equations established by the Norwegian mathematician Niles Abel in 1823. In addition, Abel's equations arise frequently in mathematical physics.

However, singular equations similar to example (8) are called the weakly-singular second-kind Volterra-type integral equations. This type of equations usually arise in scientific and engineering applications [21] like heat conduction, superfluidity and crystal growth. Recently, the weakly-singular second-kind Volterra type integral equations have been the subject of extensive analytical studies. Moreover, numerical studies have been carried out to obtain approximations of high accuracy level to the weakly-singular type of equations.

In this chapter we will focus our study on the second style of singular integral equations, namely the equations where the kernel $K(x,t)$ becomes unbounded at one or more points of singularities in its domain of definition. The equations that will be investigated are Abel's problem, generalized Abel integral equations and the weakly-singular second-kind Volterra type integral equations.

In a manner parallel to the approach used in previous chapters, we will focus our study on the techniques that will guarantee the existence of a unique solution to any singular integral equation with singularity related to the kernel $K(x,t)$ becoming unbounded at its domain of integration. We point out here that singular integral equations are in general very difficult to handle.

5.2 Abel's Problem

Abel in 1823 investigated the motion of a particle that slides down along a smooth unknown curve, in a vertical plane, under the influence of the gravitational field. It is assumed that the particle starts from rest at a point P, with vertical elevation x, slides along the unknown curve, to the lowest point O on the curve where the vertical distance is x=0. The total time of descent T from the highest point to the lowest point on the curve is given in advance, and dependent on the elevation x, hence expressed by

$$T = h(x). \tag{9}$$

Assuming that the curve of motion between the points P and O has an arclength s, then the velocity at a point Q on the curve, between P and O, is given by

$$\frac{ds}{dT} = -\sqrt{2g(x-t)}, \tag{10}$$

where t is a variable coordinate defines the vertical distance of the point Q, and g is a constant defines the acceleration of gravity. Integrating both sides of (10) gives

$$T = -\int_{O}^{P} \frac{ds}{\sqrt{2g(x-t)}}. \tag{11}$$

Setting

$$ds = u(t)dt, \tag{12}$$

and using (9) we find that the equation of motion of the sliding particle is governed by

$$f(x) = \int_{0}^{x} \frac{1}{\sqrt{x-t}} u(t)dt. \tag{13}$$

We point out that $f(x)$ is a predetermined function that depends on the elevation x and given by

$$f(x) = \sqrt{2g}\, h(x), \tag{14}$$

where g is the gravitational constant, and $h(x)$ is the time of descent from the highest point to the lowest point on the curve. The main goal of Abel's problem is to determine the unknown function $u(x)$ under the integral sign that will define the equation of the curve. Having determined $u(x)$, the equation of the smooth curve, where the particle slides along, can be easily obtained using the calculus formulas related to the arclength concepts.

It is worth mentioning that Abel's integral equation is also called Volterra integral equation of the first kind. Besides the kernel $K(x,t)$ in (13) is

$$K(x,t) = \frac{1}{\sqrt{x-t}}, \tag{15}$$

which shows that the kernel (15) is singular in that

$$K(x,t) \to \infty \quad as \quad t \to x. \tag{16}$$

The interesting Abel's problem has been approached by different methods. In the following we will employ Laplace transforms only to determine a suitable formula to solve Abel's problem (13), noting that

Laplace transforms will not be used in our approach to handle the singular equations. Taking Laplace transforms of both sides of (13) leads to

$$\begin{aligned} L[f(x)] &= L[u(x)]\, L[x^{-\frac{1}{2}}] \\ &= L[u(x)]\frac{\Gamma(1/2)}{z^{1/2}}, \end{aligned} \tag{17}$$

where Γ is the gamma function. In Appendix D, the definition of the gamma function and some of the relations related to it are given. Noting that $\Gamma(\frac{1}{2}) = \sqrt{\pi}$, the equation (17) becomes

$$L[u(x)] = \frac{z^{\frac{1}{2}}}{\sqrt{\pi}} L[f(x)], \tag{18}$$

which can be rewritten by

$$L[u(x)] = \frac{z}{\pi}\left(\sqrt{\pi} z^{-\frac{1}{2}} L[f(x)]\right). \tag{19}$$

Setting

$$h(x) = \int_0^x (x-t)^{-\frac{1}{2}} f(t)dt, \tag{20}$$

into (19) yields

$$L[u(x)] = \frac{z}{\pi} L[h(x)], \tag{21}$$

which gives

$$L[u(x)] = \frac{1}{\pi} L[h'(x)], \tag{22}$$

upon using the fact

$$L[h'(x)] = z\, L[h(x)]. \tag{23}$$

Applying L^{-1} to both sides of (22) yields the easily calculable formula

$$u(x) = \frac{1}{\pi}\frac{d}{dx}\int_0^x \frac{f(t)}{\sqrt{x-t}}dt, \tag{24}$$

that will be used for the determination of the solution. It is clear that Leibnitz rule is not applicable in (24) because the integrand is discontinuous at the interval of integration. As indicated earlier, determination of $u(x)$ will lead to the determination of the curve where the particle slides along it.

It is obvious that Abel's problem given by (13) can be solved now by using the formula (24) where the unknown function $u(x)$ has been replaced by the given function $f(x)$. One last remark concerns the use of the formula (24). The process consists of selecting the proper substitution for $(x-t)$, integrate the resulting definite integral and finally differentiate the result of the evaluation. However, for simplicity reasons we introduced Appendix B as a helpful tool needed for evaluating the integrals involved.

The procedure of using the formula (24) that determines the solution of Abel's problem (13) will be illustrated by the following examples.

Example 1. As a first example we consider the following Abel's problem

$$\pi = \int_0^x \frac{1}{\sqrt{x-t}} u(t)\, dt. \tag{25}$$

Substituting $f(x) = \pi$ in (24) yields

$$\begin{aligned} u(x) &= \frac{1}{\pi}\frac{d}{dx}\int_0^x \frac{\pi}{\sqrt{x-t}} dt, \\ &= \frac{d}{dx}\int_0^x \frac{1}{\sqrt{x-t}} dt. \end{aligned} \tag{26}$$

Setting the substitution $y = x - t$ in (26) ,we obtain

$$\begin{aligned} u(x) &= \frac{d}{dx}\left(2\sqrt{x}\right), \\ &= \frac{1}{\sqrt{x}}. \end{aligned} \tag{27}$$

Example 2. Solve the following Abel's problem

$$\frac{\pi}{2}x = \int_0^x \frac{1}{\sqrt{x-t}} u(t)\, dt. \tag{28}$$

Substituting $f(x) = \frac{\pi}{2}x$ in (24) gives

$$\begin{aligned} u(x) &= \frac{1}{\pi}\frac{d}{dx}\int_0^x \frac{\frac{\pi}{2}t}{\sqrt{x-t}} dt, \\ &= \frac{1}{2}\frac{d}{dx}\int_0^x \frac{t}{\sqrt{x-t}} dt. \end{aligned} \tag{29}$$

Using integration by substitution, where we set $y = x - t$, or by using Appendix B, we obtain

$$\begin{aligned} u(x) &= \frac{1}{2}\frac{d}{dx}\left(\frac{4}{3}x^{\frac{3}{2}}\right), \\ &= x^{\frac{1}{2}}. \end{aligned} \tag{30}$$

Example 3. As a third example we consider the following Abel's problem

$$2\sqrt{x} = \int_0^x \frac{1}{\sqrt{x-t}}u(t)\,dt, \tag{31}$$

Substituting $f(x) = 2\sqrt{x}$ in (24) we find

$$u(x) = \frac{2}{\pi}\frac{d}{dx}\int_0^x \frac{\sqrt{t}}{\sqrt{x-t}}dt. \tag{32}$$

The integral at the right hand side of (32) can be evaluated by using integration by substitution, where in this case we set the substitution

$$t = x\sin^2\theta, \tag{33}$$

so that

$$\sqrt{x-t} = \sqrt{x}\cos\theta, \tag{34}$$

and

$$dt = 2x\sin\theta\cos\theta d\theta. \tag{35}$$

Substituting (33) – (35) in (32) we obtain

$$\begin{aligned} u(x) &= \frac{4}{\pi}\frac{d}{dx}\left(x\int_0^{\pi/2}\sin^2\theta d\theta\right), \\ &= \frac{4}{\pi}\frac{d}{dx}\left(x\left[\frac{1}{2}\theta - \frac{1}{4}\sin(2\theta)\right]_0^{\pi/2}\right), \\ &= 1. \end{aligned} \tag{36}$$

It should be noted that Eq. (32) can be solved directly, where the integral at the right hand side may be evaluated directly by using Appendix B.

5.2.1 The Generalized Abel's Integral Equation

It is important here to note that Abel introduced the more general singular integral equation

$$f(x) = \int_0^x \frac{1}{(x-t)^\alpha} u(t)dt, \quad 0 < \alpha < 1, \tag{37}$$

known as the *Generalized Abel's integral equation.* It can be easily seen that Abel's problem discussed above is a special case of the *generalized equation* where $\alpha = \frac{1}{2}$. To determine a practical formula for the solution $u(x)$ of (37), and hence for the Abel's problem, we simply use the Laplace transform in a similar manner to that used above. As noted before, the Laplace transform will be used for the derivation of the proper formula, but will not be used in handling the equations. Taking Laplace transforms to both sides of Eq. (37) yields

$$\begin{aligned} L[f(x)] &= L[u(x)]\,L[x^{-\alpha}] \\ &= L[u(x)]\frac{\Gamma(1-\alpha)}{z^{1-\alpha}}, \end{aligned} \tag{38}$$

where Γ is the gamma function. The equation (38) can be written as

$$L[u(x)] = \frac{z}{\Gamma(\alpha)\Gamma(1-\alpha)}\Gamma(\alpha)z^{-\alpha}L[f(x)], \tag{39}$$

or equivalently

$$L[u(x)] = \frac{z}{\Gamma(\alpha)\Gamma(1-\alpha)}L[g(x)], \tag{40}$$

where

$$g(x) = \int_0^x (x-t)^{\alpha-1}f(t)dt. \tag{41}$$

Accordingly, Eq. (40) can be written as

$$L[u(x)] = \frac{\sin(\alpha\pi)}{\pi}\,L[g'(x)], \tag{42}$$

upon using the identities

$$L[g'(x)] = z\,L[g(x)], \tag{43}$$

and

$$\Gamma(\alpha)\Gamma(1-\alpha) = \frac{\pi}{\sin(\alpha\pi)}, \tag{44}$$

from Laplace transforms and Appendix D respectively. Applying L^{-1} to both sides of (42) yields the easily calcualable formula for determining the solution

$$u(x) = \frac{\sin(\alpha\pi)}{\pi}\frac{d}{dx}\int_0^x \frac{f(t)}{(x-t)^{1-\alpha}}dt, \quad 0 < \alpha < 1. \tag{45}$$

Recall that $f(x)$ is differentiable, we therefore can derive from (45) a more suitable formula that will support our computational purposes. To determine this formula, we first integrate the integral at the right hand side of (45) by parts where we obtain

$$\begin{aligned}\int_0^x \frac{f(t)}{(x-t)^{1-\alpha}}dt &= -\frac{1}{\alpha}\left[f(t)(x-t)^{\alpha}\right]_0^x + \frac{1}{\alpha}\int_0^x (x-t)^{\alpha} f'(t)dt, \\ &= \frac{1}{\alpha}f(0)x^{\alpha} + \frac{1}{\alpha}\int_0^x (x-t)^{\alpha} f'(t)dt.\end{aligned} \tag{46}$$

Differentiating both sides of (46), noting that Leibnitz rule should be used in differentiating the integral at the right hand side, yields

$$\frac{d}{dx}\int_0^x \frac{f(t)}{(x-t)^{1-\alpha}}dt = \frac{f(0)}{x^{1-\alpha}} + \int_0^x \frac{f'(t)}{(x-t)^{1-\alpha}}dt. \tag{47}$$

Substituting (47) into (45) yields the desired formula given by

$$u(x) = \frac{\sin(\alpha\pi)}{\pi}\left(\frac{f(0)}{x^{1-\alpha}} + \int_0^x \frac{f'(t)}{(x-t)^{1-\alpha}}dt\right), \quad 0 < \alpha < 1, \tag{48}$$

that will be used to determine the solution of the generalized Abel's equation and consequently, of the standard Abel's problem as well.

The following illustrative examples explain how we can use the formula (48) in solving Abel's equations.

Example 4. Solve the following Abel's problem

$$\pi x = \int_0^x \frac{1}{\sqrt{x-t}}u(t)\,dt. \tag{49}$$

In this example $f(x) = \pi x$, hence $f(0) = 0$ and $f'(x) = \pi$. Besides, $\alpha = \frac{1}{2}$, hence $\sin(\alpha\pi) = 1$. Using the formula (48) we obtain

$$\begin{aligned} u(x) &= \frac{1}{\pi}\int_0^x \frac{\pi}{\sqrt{x-t}}dt, \\ &= 2\sqrt{x}, \end{aligned} \tag{50}$$

by using Appendix B.

Example 5. Solve the following Abel's problem

$$4\sqrt{x} = \int_0^x \frac{1}{\sqrt{x-t}}u(t)\,dt. \tag{51}$$

In this example $f(x) = 4\sqrt{x}$, hence $f(0) = 0$ and $f'(x) = \frac{2}{\sqrt{x}}$. Besides, $\alpha = \frac{1}{2}$, hence $\sin(\alpha\pi) = 1$. Using the formula (48) we find

$$u(x) = \frac{1}{\pi}\int_0^x \frac{2}{\sqrt{t}\sqrt{x-t}}dt. \tag{52}$$

Using the substitution

$$t = x\sin^2\theta, \tag{53}$$

in (52) gives the exact solution

$$u(x) = 2. \tag{54}$$

It should be noted that the integral at the right hand side of Eq. (52) can also be evaluated by using regular table of integrals.

Example 6. Find an approximate solution to the following Abel's problem

$$\sinh x = \int_0^x \frac{1}{\sqrt{x-t}}u(t)\,dt. \tag{55}$$

In this example $f(x) = \sinh x$, hence $f(0) = 0$ and $f'(x) = \cosh x$. Using the formula (48) we find

$$u(x) = \frac{1}{\pi}\int_0^x \frac{\cosh t}{\sqrt{x-t}}dt. \tag{56}$$

An approximate solution can be found by considering $\cosh x \approx 1 + \frac{x^2}{2!}$ for small x. Consequently, we have

$$u(x) \approx \frac{1}{\pi}\int_0^x \frac{1+\frac{t^2}{2}}{\sqrt{x-t}}dt, \tag{57}$$

which gives

$$u(x) \approx \frac{2}{15\pi}\sqrt{x}(15+4x^2), \quad \text{for small } x, \tag{58}$$

by integrating by substitution or by using Appendix B.

For simplicity reasons, we recommend the use of Appendix B for evaluating the resulting integrals in the following exercises.

Exercises 5.2

Solve the following Abel's integral equations:

1. $\pi(x+1) = \int_0^x \frac{1}{\sqrt{x-t}}u(t)\,dt.$

2. $\frac{\pi}{2}(x^2-x) = \int_0^x \frac{1}{\sqrt{x-t}}u(t)\,dt.$

3. $x^2+x+1 = \int_0^x \frac{1}{\sqrt{x-t}}u(t)\,dt.$

4. $\frac{3\pi}{8}x^2 = \int_0^x \frac{1}{\sqrt{x-t}}u(t)\,dt.$

5. $\frac{4}{3}x^{\frac{3}{2}} = \int_0^x \frac{1}{\sqrt{x-t}}u(t)\,dt.$

6. $\frac{8}{15}x^{\frac{5}{2}} = \int_0^x \frac{1}{\sqrt{x-t}}u(t)\,dt.$

7. $x^3 = \int_0^x \frac{1}{\sqrt{x-t}}u(t)\,dt.$

8. $x^4 = \int_0^x \frac{1}{\sqrt{x-t}}u(t)\,dt.$

9. $x+x^3 = \int_0^x \frac{1}{\sqrt{x-t}}u(t)\,dt.$

10. $\sin x = \int_0^x \frac{1}{\sqrt{x-t}}u(t)\,dt.$

5.3 The Weakly-Singular Volterra Equations

As indicated earlier, the weakly-singular second-kind Volterra-type integral equations of the second kind, given by

$$u(x) = g(x) + \int_0^x \frac{\beta}{\sqrt{x-t}} u(t)dt, \quad x \in [0, T], \tag{59}$$

appear frequently in many mathematical physics and chemistry applications such as heat conduction, crystal growth and electrochemistry [21]. It is to be noted that β is a constant and $T = 1, 2,$ or 3 depending on the science model under discussion. It is also assumed that the function $g(x)$ is sufficiently smooth so that a unique solution to (59) is guaranteed. The equation (59) falls under the category of singular equations with singular kernel $K(x,t) = \frac{1}{\sqrt{x-t}}$.

A considerable amount of work has been carried recently on these models to determine its exact solutions or to achieve numerical approximations of high degree of accuracy. Collocation type methods were employed to address this type of equations to achieve numerical approximations.

In this section we will base our discussion on the decomposition method [25] that was introduced in the preceding chapters. We will show that this technique is an effective and powerful tool to handle this type of singular equations analytically and numerically. The method has been discussed extensively and need not be introduced in details here. In the following we outline a brief framework of the method. To determine the solution $u(x)$ of (59) we usually use the decomposition

$$u(x) = \sum_{n=0}^{\infty} u_n(x), \tag{60}$$

into both sides of (59) to obtain

$$\sum_{n=0}^{\infty} u_n(x) = g(x) + \int_0^x \frac{\beta}{\sqrt{x-t}} \left(\sum_{n=0}^{\infty} u_n(t) \right) dt, \quad x \in [0, T]. \tag{61}$$

The components $u_0(x), u_1(x), u_2(x), \ldots$ are immediately determined upon applying the following recurrent relations

$$\begin{cases} u_0(x) &= g(x) \\ u_1(x) &= \displaystyle\int_0^x \frac{\beta}{\sqrt{x-t}} u_0(t)dt, \\ u_2(x) &= \displaystyle\int_0^x \frac{\beta}{\sqrt{x-t}} u_1(t)dt, \\ & \vdots \end{cases} \tag{62}$$

Having determined the components $u_0(x), u_1(x), u_2(x), ...,$ the solution $u(x)$ of (59) will be easily obtained in the form of a rapid convergent power series by substituting the derived components in (60). It is convenient to use Appendix B to evaluate the integrals in (62).

It is important to note that the phenomena of the self-cancelling noise terms, where like terms with opposite signs appear in specific problems, should be observed here between the components $u_0(x)$ and $u_1(x)$. As mentioned earlier, the appearance of these terms usually speeds the convergence of the solution and normally minimizes the size of the computational work. It is sometimes convenient to use the modified decomposition method as discussed above. For illustration purposes, we discuss the following examples.

Example 1. We first consider the weakly-singular second-type Volterra integral equation

$$u(x) = \sqrt{x} + \frac{1}{2}\pi x - \int_0^x \frac{1}{\sqrt{x-t}} u(t)\, dt, \quad I = [0,2]. \tag{63}$$

Using the recurrent algorithm we set

$$u_0(x) = \sqrt{x} + \frac{1}{2}\pi x, \tag{64}$$

which gives

$$u_1(x) = -\int_0^x \frac{\sqrt{t} + \frac{1}{2}\pi t}{\sqrt{x-t}}\, dt. \tag{65}$$

The transformation

$$t = x \sin^2\theta, \tag{66}$$

carries (65) into

$$\begin{aligned} u_1(x) &= -\int_0^{\pi/2} \left(x \sin^2\theta + \pi x^{3/2} \sin^3\theta \right) d\theta, \\ &= -\frac{1}{2}\pi x - \frac{2}{3}\pi x^{3/2}. \end{aligned} \tag{67}$$

The result (67) can be obtained directly from (65) by using Appendix B. Observing the appearance of the terms $\frac{1}{2}\pi x$ and $-\frac{1}{2}\pi x$ between the components $u_0(x)$ and $u_1(x)$, and verifying that the non-cancelled term in $u_0(x)$ justifies the equation (63) yields

$$u(x) = \sqrt{x}, \tag{68}$$

the exact solution of the equation

It can be shown that it is possible to obtain the exact solution (68) by using the modified decomposition method. This can be done by splitting the nonhomogeneous part $g(x)$ into two parts. Accordingly, we set

$$u_0(x) = \sqrt{x}, \tag{69}$$

so that

$$u_1(x) = \frac{1}{2}\pi x - \int_0^x \frac{\sqrt{t}}{\sqrt{x-t}}\, dt, \tag{70}$$

which gives

$$u_1(x) = 0. \tag{71}$$

Consequently, other components will vanish, and the exact solution (68) follows immediately.

Example 2. As a second example we consider the weakly-singular second-kind Volterra integral equation

$$u(x) = x + \frac{4}{3}x^{3/2} - \int_0^x \frac{1}{\sqrt{x-t}} u(t)\, dt, \quad I = [0, 2]. \tag{72}$$

Proceeding as before we set

$$u_0(x) = x + \frac{4}{3}x^{3/2}, \tag{73}$$

which gives

$$u_1(x) = -\int_0^x \frac{t+\frac{4}{3}t^{3/2}}{\sqrt{x-t}}\,dt. \tag{74}$$

Using the transformation (66) we obtain

$$\begin{aligned} u_1(x) &= -\int_0^{\pi/2}\left(2x\sin^3\theta + \frac{8}{3}x^2\sin^4\theta\right)d\theta, \\ &= -\frac{4}{3}x^{3/2} - \frac{1}{2}\pi x^2. \end{aligned} \tag{75}$$

The resulting value of $u_1(x)$ in (75) can be determined directly by using Appendix B when evaluating (74).

Cancelling the noise terms between the components $u_0(x)$ and $u_1(x)$ in (73) and (75), and verifying that the remaining term in $u_0(x)$ satisfies the equation (72) gives the exact solution

$$u(x) = x. \tag{76}$$

As discussed in example 1, we can obtain the exact solution (76) by using the modified decomposition method. We leave it as an exercise to the reader.

Example 3. As a final example we consider the weakly-singular second-type Volterra integral equation

$$u(x) = 2\sqrt{x} - \int_0^x \frac{1}{\sqrt{x-t}}u(t)\,dt, \quad I = [0,2]. \tag{77}$$

Following the discussion in the previous examples we set

$$u_0(x) = 2\sqrt{x}, \tag{78}$$

which gives

$$u_1(x) = -2\int_0^x \frac{\sqrt{t}}{\sqrt{x-t}}\,dt, \tag{79}$$

and by using the transformation (66) or Appendix B we obtain

$$u_1(x) = -\pi x. \tag{80}$$

It is clear that the self-cancelling noise terms did not appear in the components $u_0(x)$ and $u_1(x)$, therefore we continue to determine as many components as we desire. Consequently, we find

$$u_2(x) = \int_0^x \frac{\pi t}{\sqrt{x-t}}\, dt, \tag{81}$$

which gives

$$u_2(x) = \frac{4}{3}\pi x^{3/2}, \tag{82}$$

by using Appendix B. The component $u_3(x)$ can be easily determined in a similar manner, hence we find

$$u_3(x) = -\frac{4}{3}\int_0^x \frac{\pi t^{3/2}}{\sqrt{x-t}}\, dt, \tag{83}$$

which gives

$$u_3(x) = -\frac{1}{2}\pi^2 x^2. \tag{84}$$

Combining (78), (80), (82), and (84), the solution $u(x)$ in a series form

$$u(x) = 2\sqrt{x} - \pi x + \frac{4}{3}\pi x^{3/2} - \frac{1}{2}\pi^2 x^2 + \cdots, \tag{85}$$

is readily obtained. The result obtained in (85) can be expressed as

$$u(x) = 2\sqrt{x} - \pi x + O(x^{\frac{3}{2}}), \text{ as } x \to 0. \tag{86}$$

It is to be noted that in this example we determined four components of the series solution. Other components can be obtained in a similar fashion to increase the degree of accuracy for numerical purposes. However, the exact solution of (77) is given by

$$u(x) = 1 - e^{\pi x}\operatorname{erfc}(\sqrt{\pi x}), \tag{87}$$

where $erfc$ is the complementary error function normally used in probability topics. We can easily show that the exact solution given by (87) produces the series solution of (85) upon using the asymptotic expansion of the complementary error function given in Appendix D. The definitions of the error function and the complementary error function and the related properties can be found in Appendix D.

Exercises 5.3

Use the decomposition method or the modified decomposition method to solve the following weakly-singular second-kind Volterra integral equations:

1. $u(x) = \sqrt{x} - \pi x + 2\int_0^x \frac{1}{\sqrt{x-t}} u(t)\, dt, \quad I = [0, 2].$
2. $u(x) = x^{\frac{3}{2}} + \frac{3}{8}\pi x^2 - \int_0^x \frac{1}{\sqrt{x-t}} u(t)\, dt, \quad I = [0, 2].$
3. $u(x) = \frac{1}{2} - \sqrt{x} + \int_0^x \frac{1}{\sqrt{x-t}} u(t)\, dt, \quad I = [0, 2].$
4. $u(x) = \sqrt{x} - \frac{1}{2}\pi x + \int_0^x \frac{1}{\sqrt{x-t}} u(t)\, dt, \quad I = [0, 2].$
5. $u(x) = x^{\frac{5}{2}} - \frac{5}{16}\pi x^3 + \int_0^x \frac{1}{\sqrt{x-t}} u(t)\, dt, \quad I = [0, 2]$.
6. $u(x) = x^3 + \frac{32}{33} x^{\frac{7}{2}} - \int_0^x \frac{1}{\sqrt{x-t}} u(t)\, dt, \quad I = [0, 2].$
7. $u(x) = 1 + x - 2\sqrt{x} - \frac{4}{3} x^{\frac{3}{2}} + \int_0^x \frac{1}{\sqrt{x-t}} u(t)\, dt, \quad I = [0, 2].$
8. $u(x) = 1 + 2\sqrt{x} - \int_0^x \frac{1}{\sqrt{x-t}} u(t)\, dt, \quad I = [0, 1].$
9. $u(x) = x^2 + \frac{16}{15} x^{\frac{5}{2}} - \int_0^x \frac{1}{\sqrt{x-t}} u(t)\, dt, \quad I = [0, 1].$
10. $u(x) = \frac{2}{\pi}\sqrt{x} + \frac{15}{16} x^2 - x - x^{\frac{5}{2}} + \int_0^x \frac{1}{\sqrt{x-t}} u(t)\, dt, \quad I = [0, 2].$

Chapter 6

Nonlinear Integral Equations

6.1 Definitions

So far in this text we have been mainly concerned with studying different methods for solving linear integral equations of the second kind. We pointed out earlier that nonlinear integral equations yield a considerable amount of difficulties. However, with the recent methods developed, it seems reasonable to present some reliable and powerful techniques that will make the study of specific cases of nonlinear integral equations successful and valuable. In general, the solution of the nonlinear integral equation is not unique [10], [17] and [18]. However, the existence of a unique solution of nonlinear integral equations with specific conditions is possible, but uniqueness concept will not be discussed in this text as mentioned earlier.

The purpose of this chapter is to introduce reliable and easily computable techniques for solving specific cases of nonlinear integral equations. As indicated in Chapter 1, given $F(u(t))$ a nonlinear function in $u(t)$, integral equations of the form

$$u(x) = f(x) + \lambda \int_a^b K(x,t)\, F(u(t))dt, \tag{1}$$

and

$$u(x) = f(x) + \lambda \int_0^x K(x,t)\, F(u(t))dt, \tag{2}$$

are called *nonlinear* Fredholm integral equations and *nonlinear* Volterra integral equations respectively. The function $F(u(t))$ is nonlinear in $u(t)$ such as $u^2(t)$, $u^3(t)$, $e^{u(t)}$, and λ is a parameter. However, in this text, we will restrict our discussion to the case where $F(u(t)) = u^n(t)$, $n \geq 2$, whereas other nonlinear integral equations that involve nonlinear terms other than $u^n(t)$ can be handled in a very similar manner. The following are examples of the *nonlinear* integral equations:

$$u(x) = 1 + \lambda \int_0^1 u^2(t)dt, \tag{3}$$

$$u(x) = x + \int_0^1 xtu^3(t)dt, \tag{4}$$

$$u(x) = x - \frac{1}{4}x^4 + \int_0^x tu^2(t)dt, \tag{5}$$

$$u(x) = 2x + \frac{1}{6}x^5 - \int_0^x tu^3(t)\, dt, \tag{6}$$

where examples (3) – (4) and (5) – (6) are nonlinear Fredholm and nonlinear Volterra integral equations respectively.

6.2 Nonlinear Fredholm Integral Equations

In this section we will discuss the most successful methods for solving nonlinear Fredholm integral equations (1). It has been concluded that the *direct computation method* proved to be reliable in that it handled successfully the linear Fredholm integral equations and the Fredholm integro-differential equations in Chapters 2 and 4 respectively. Based on this conclusion, the direct computation method will be implemented here to provide the exact or (closed form) solution as will be discussed later.

In addition, the *decomposition method* proved to be an elegant tool in handling linear and nonlinear equations as well. Accordingly, it is useful to use this method here to obtain the solution in the form of a power series. However, one important fact concerning the decomposition method, in handling the nonlinear problems, is that it requires the use of the so called Adomian polynomials that will represent the involved nonlinear function $u^n(t)$, $n \geq 2$ that appears under the integral sign. The scheme that will generate these polynomials will be introduced later.

In the following we will discuss the steps needed to implement these methods.

6.2.1 The Direct Computation Method

As stated before we will focus our study on the nonlinear Fredholm integral equations of the form

$$u(x) = f(x) + \lambda \int_a^b K(x,t)\, u^n(t)dt, \tag{7}$$

where the kernel $K(x,t)$ will be assumed a separable kernel. Without less of generality, we may consider the kernel $K(x,t)$ to be expressed by

$$K(x,t) = g(x)h(t). \tag{8}$$

Consequently, we rewrite the equation (7) as

$$u(x) = f(x) + \lambda g(x) \int_a^b h(t)\, u^n(t)dt, \tag{9}$$

We can easily observe that the definite integral in the right hand side of (9) depends only on the variable t. Therefore, we will follow the approach usually used in the *direct computation method*, hence we set

$$\alpha = \int_a^b h(t)\, u^n(t)dt, \tag{10}$$

where the constant α represents the numerical value of the integral. Accordingly, we may rewrite (9) as

$$u(x) = f(x) + \lambda \alpha g(x). \tag{11}$$

Substituting $u(x)$ from (11) into (10), and integrating the easily computable integral yield the numerical value of the constant α. The exact solution $u(x)$ is readily determined upon substituting the obtained value of α in (11).

We point out that the derived solution $u(x)$ in (11) depends on the parameter λ. Accordingly,it is normal to discuss all possible values of λ that will define real solutions for $u(x)$. As a result, two related phenomenas , termed as the *bifurcation point* and the *singular point*, may appear. These phenomenas have been introduced by [20] and others. For simplicity reasons, the direct computation method and the phenomenas of the *bifurcation point* and the *singular point* of the nonlinear integral equation will be illustrated by the following examples.

Example 1. We consider the nonlinear Fredholm integral equation

$$u(x) = 2 + \lambda \int_0^1 u^2(t)\, dt, \tag{12}$$

Setting

$$\alpha = \int_0^1 u^2(t)\, dt, \tag{13}$$

carries (12) into

$$u(x) = 2 + \lambda\alpha. \tag{14}$$

Substituting (14) into (13) yields

$$\alpha = \int_0^1 (2 + \lambda\alpha)^2\, dt, \tag{15}$$

which gives

$$\alpha = (2 + \lambda\alpha)^2, \tag{16}$$

or equivalently

$$\lambda^2\alpha^2 + (4\lambda - 1)\alpha + 4 = 0. \tag{17}$$

Solving the quadratic equation (17) for α gives

$$\alpha = \frac{(1 - 4\lambda) \pm \sqrt{1 - 8\lambda}}{2\lambda^2}, \tag{18}$$

so that substituting (18) into (14) yields

$$u(x) = \frac{1 \pm \sqrt{1-8\lambda}}{2\lambda}. \tag{19}$$

The following phenomenas, related to all possible values of λ, were investigated by [20] and others, hence it should be observed:

(i) For $\lambda = 0$, using (12) we obtain $u(x) = 1$, but using (19) we find that $u(x)$ is infinite. For this reason, the point $\lambda = 0$ is called *a singular point* of the equation (12).

(ii) For $\lambda < \frac{1}{8}$, the equation (12) has two real solutions. It is clear in this case that the solution is not unique. This is normal for nonlinear integral equations.

(iii) For $\lambda = \frac{1}{8}$, the equation (12) has one real solution and the point $\lambda = \frac{1}{8}$ is called a *bifurcation point.* The real solution in this case is $u(x) = 4$.

Example 2. We next consider the nonlinear Fredholm integral equation

$$u(x) = \frac{7}{8}x + \frac{1}{2}\int_0^1 xtu^2(t)\,dt. \tag{20}$$

Setting

$$\alpha = \int_0^1 tu^2(t)\,dt, \tag{21}$$

carries (20) into

$$u(x) = (\frac{7}{8} + \frac{1}{2}\alpha)x. \tag{22}$$

Substituting (22) into the equation (21) gives

$$\alpha = \int_0^1 t\left(\frac{7}{8} + \frac{1}{2}\alpha\right)^2 t^2 dt, \tag{23}$$

which gives

$$\alpha = \frac{1}{4}\left(\frac{7}{8} + \frac{1}{2}\alpha\right)^2, \tag{24}$$

or equivalently

$$(4\alpha - 1)(4\alpha - 49) = 0, \tag{25}$$

so that

$$\alpha = \frac{1}{4}, \frac{49}{4}. \tag{26}$$

Accordingly, two real solutions given by

$$u(x) = x, 7x, \tag{27}$$

are obtained upon using (26) into (22).

Exercises 6.2.1

In exercises 1 – 5, use the *Direct Computation Method* to solve the given nonlinear integral equations. Also find the *singular point* and the *bifurcation point* of each equation

1. $u(x) = 1 + \frac{1}{2}\lambda \int_0^1 u^2(t)\,dt$

2. $u(x) = 1 - \lambda \int_0^1 u^2(t)\,dt$

3. $u(x) = 1 + \lambda \int_0^1 tu^2(t)\,dt$

4. $u(x) = 1 + \lambda \int_0^1 t^2u^2(t)\,dt$

5. $u(x) = 1 + \lambda \int_0^1 t^3u^2(t)\,dt$

In exercises 6 – 12, use the *Direct Computation Method* to solve the following nonlinear integral equations

6. $u(x) = 2 - \frac{4}{3}x + \int_0^1 xt^2u^2(t)\,dt$

7. $u(x) = \sin x - \frac{\pi}{8} + \frac{1}{2}\int_0^{\pi/2} u^2(t)\,dt$

8. $u(x) = \cos x - \frac{\pi}{8} + \frac{1}{2}\int_0^{\pi/2} u^2(t)\,dt$

9. $u(x) = x - \frac{1}{8} + \frac{1}{2}\int_0^1 tu^2(t)\,dt$

10. $u(x) = x^2 - \frac{1}{10} + \frac{1}{2}\int_0^1 u^2(t)\,dt$

11. $u(x) = x - \frac{5}{6} + \int_0^1 \left(u(t) + u^2(t)\right)\,dt$

12. $u(x) = x - 1 + \frac{3}{4}\int_0^1 \left(2t + u^2(t)\right)\,dt$

6.2.2 The Decomposition Method

The Adomian decomposition method introduces a reliable analysis to handle the nonlinear integral equations. The method provides a rapidly convergent series solution without using any restrictive assumptions such as linearization. The linear term $u(x)$ in the equation is usually expressed in a power series in a similar manner to that discussed before for linear integral equations. However, an essential scheme is required for representing the nonlinear term $u^n(x)$ involved in the equation. Moreover, the decomposition method does not investigate the existence and the uniqueness of the solution of the problem.

In the following , the decomposition method will be fully discussed for nonlinear equations [1] and [2]. For simplicity reasons, we consider the simple form of nonlinear Fredholm integral equation

$$u(x) = f(x) + \lambda \int_a^b K(x,t)\, u^n(t)dt, \tag{28}$$

where other forms of nonlinearity of $F(u(t))$ can be handled in a parallel manner. The solution $u(t)$ of (28) can be represented normally by the decomposition series

$$u(x) = \sum_{n=0}^{\infty} u_n(x), \tag{29}$$

where the components $u_n(x)$, $n \geq 0$ can be computed in a recursive manner as discussed before. However, as stated above the nonlinear term $u^n(t)$ of the equation (28) should be represented, using a distinct scheme, by the so called Adomian polynomials $A_n(t)$. The simple and practical scheme that will generate these polynomials begins by assuming that the nonlinear term $u^n(t)$ under the integral sign in (28) will be equated to the polynomial series

$$u^n(t) = \sum_{n=0}^{\infty} A_n(t), \tag{30}$$

where the $A_n(t)$ are the so called Adomian polynomials. It was formally proved by [2] that the Adomian polynomials can be completely determined by using the following scheme

$$\left\{\begin{aligned} A_0 &= F(u_0), \\ A_1 &= u_1\, F'(u_0), \\ A_2 &= u_2\, F'(u_0) + \tfrac{u_1^2}{2!}\, F''(u_0), \\ A_3 &= u_3\, F'(u_0) + u_1 u_2\, F''(u_0) + \tfrac{u_1^3}{3!}\, F'''(u_0), \\ A_4 &= u_4\, F'(u_0) + \left(\tfrac{1}{2!}u_2^2 + u_1 u_3\right) F''(u_0) \\ &\quad + \tfrac{u_1^2 u_2}{2!}\, F'''(u_0) + \tfrac{1}{4!}u_1^4 F^{(iv)}(u_0), \end{aligned}\right. \tag{31}$$

where $F(u(t))$ is the nonlinear function, and in this specific equation it is given by $F(u(t)) = u^n(t)$. In addition we point out that A_0 depends only on u_0, A_1 depends only on u_0 and u_1, A_2 depends only on u_0, u_1 and u_2, etc. It is to be noted that the sum of the subscripts of each term of A_n is equal to n. More details about the derivation of Adomian polynomials can be found in [1] and [2]. A simplified technique to derive Adomian polynomials is under discussion .

It is remarked before that in this section, for simplicity reasons, we will discuss nonlinear Fredholm integral equation (28) where the nonlinear term is of the form $F(u(t)) = u^n(t)$ only.

In the following we explain how we can use the scheme given by (31) to define Adomian polynomials:

(i) Consider the nonlinear function

$$F(u) = u^2(x), \tag{32}$$

then

$$\begin{cases} A_0 &= u_0^2, \\ A_1 &= 2u_0u_1 \\ A_2 &= 2u_0u_2 + u_1^2 \\ A_3 &= 2u_0u_3 + 2u_1u_2 \\ \vdots \end{cases} \tag{33}$$

We can easily observe from (33) that in the polynomial A_2 the sum of the subscripts of each term of the two terms $2u_0u_2$ and u_1u_1 is equal to the subscript of A_2. The same fact holds for other polynomials.

(ii) For the nonlinear function

$$F(u) = u^3 \tag{34}$$

we find

$$\begin{cases} A_0 &= u_0^3 \\ A_1 &= 3u_0^2u_1 \\ A_2 &= 3u_0^2u_2 + 3u_0u_1^2 \\ A_3 &= 3u_0^2u_3 + 6u_0u_1u_2 + u_1^3 \\ \vdots \end{cases} \tag{35}$$

(iii) For the nonlinear function

$$F(u) = u^4 \tag{36}$$

we find

$$\begin{cases} A_0 &= u_0^4 \\ A_1 &= 4u_0^3u_1 \\ A_2 &= 4u_0^3u_2 + 6u_0^2u_1^2 \\ A_3 &= 4u_0^3u_3 + 12u_0^2u_1u_2 + 4u_0u_1^3 \\ \vdots \end{cases} \tag{37}$$

(iv) For the nonlinear function

$$F(u) = e^u \tag{38}$$

we use the scheme (31) to generate the Adomian polynomials

$$\begin{cases} A_0 &= e^{u_0} \\ A_1 &= u_1 e^{u_0} \\ A_2 &= \left(\dfrac{u_1^2}{2} + u_2\right) e^{u_0} \\ A_3 &= \left(\dfrac{u_1^3}{6} + u_1 u_2 + u_3\right) e^{u_0} \\ \vdots \end{cases} \tag{39}$$

The last example, where the nonlinear function $F(u) = e^u$, has been introduced for further studies beyond the scope of this text. This example is presented here for illustration purposes only. As stated before, $F(u)$ may have other nonlinear forms such as $\sin(u)$, $\cos(u)$. However, our concern in this text will be on nonlinear functions of the form $F(u) = u^n$. We point out that for , $n \geq 3$, the decomposition method is easier to use than the direct computation method. In the latter case, an algebraic equation of higher degree is obtained.

We now return to the main goal of our discussion to determine the components $u_0(x), u_1(x), u_2(x), \ldots$ of the solution $u(x)$. This can be done by substituting the decomposition (29) that represents the linear term $u(x)$, and the decomposition (30) that represents the nonlinear term $u^n(x)$ into (28). Hence we obtain

$$\sum_{n=0}^{\infty} u_n(x) = f(x) + \lambda \int_a^b K(x,t) \left(\sum_{n=0}^{\infty} A_n(t)\right) dt, \tag{40}$$

or simply

$$\begin{aligned} u_0(x) &+ u_1(x) + u_2(x) + \cdots \\ &= f(x) + \lambda \int_a^b K(x,t) \left[A_0(t) + A_1(t) + A_2(t) + \cdots\right] dt. \end{aligned} \tag{41}$$

The components $u_0(x), u_1(x), u_2(x), \ldots$ are completely determined by using the recurrent relationship

$$\begin{cases} u_0(x) &= f(x), \\ u_1(x) &= \lambda \int_a^b K(x,t) A_0(t)dt, \\ u_2(x) &= \lambda \int_a^b K(x,t) A_1(t)dt, \\ \vdots \\ u_{n+1}(x) &= \lambda \int_a^b K(x,t) A_n(t)dt, \quad n \geq 0. \end{cases} \tag{42}$$

Consequently the solution of (28) in a series form is immediately determined. As indicated earlier, the series obtained may yield the exact solution in a closed form, or a truncated series $\sum_{n=1}^{k} u_n(x)$ may be used if a numerical approximation is desired. It is worthnoting that the convergence question of the method was addressed by [2] and extensively studied by [8].

Before we give a clear view of the method that handles the occurence of the nonlinear term $u^n(x)$, it is useful to discuss the following remarks:

(i) Even though the decomposition method gives only one solution for each nonlinear equation, this does not indicate the uniqueness of the solution of the nonlinear integral equations. We have seen that two real solutions were obtained for examples discussed above by using the direct computation method, and a unique solution is determined under specific conditions only. This is consistent with the fact that the decomposition method does not address the existence and the uniqueness concepts.

(ii) The modified decomposition method, that was introduced before and the criteria of the self-cancelling noise terms can be implemented here to speed the process of obtaining the solution.

(iii) It is important to emphasize that Adomian polynomials A_n can be calculated for complicated nonlinearities of $F(u)$.

In the following we will outline a brief framework of the *modified decomposition method*. Splitting the nonhomogeneous part $f(x)$ into two parts $f_1(x)$ and $f_2(x)$ enables us to follow the scheme given by

$$\begin{cases} u_0(x) &= f_1(x), \\ u_1(t) &= f_2(x) + \lambda \int_a^b K(x,t)A_0(t)dt, \\ u_2(t) &= \lambda \int_a^b K(x,t)A_1(t)dt, \\ \vdots & \\ u_{n+1}(x) &= \lambda \int_a^b K(x,t)A_n(t)dt, \quad n \geq 0. \end{cases} \tag{43}$$

Having determined the components $u_0(x), u_1(x), u_2(x), \ldots$ leads to the solution in a series form upon using (29).

The decomposition method and the modified decomposition method will be illustrated by the following examples. For comparison reasons, we discuss Examples 1 and 2 above that were solved before by using the direct computation method.

Example 1. We consider the nonlinear Fredholm integral equation

$$u(x) = 2 + \lambda \int_0^1 u^2(t)\,dt, \quad \lambda \leq \frac{1}{8}. \tag{44}$$

In this example we have

$$F(u) = u^2(x), \tag{45}$$

which generates the following polynomials

$$\begin{cases} A_0 &= u_0^2, \\ A_1 &= 2u_0u_1 \\ A_2 &= 2u_0u_2 + u_1^2 \\ A_3 &= 2u_0u_3 + 2u_1u_2 \\ \vdots & \end{cases} \tag{46}$$

Using the recursive algorithm (42) we find

$$u_0(x) = 2, \tag{47}$$

$$\begin{aligned} u_1(x) &= \lambda \int_0^1 A_0(t)dt, \\ &= 4\lambda, \end{aligned} \tag{48}$$

$$\begin{aligned} u_2(x) &= \lambda \int_0^1 A_1(t)dt, \\ &= 16\lambda^2, \end{aligned} \tag{49}$$

$$\begin{aligned} u_3(x) &= \lambda \int_0^1 A_2(t)dt, \\ &= 80\lambda^3, \end{aligned} \tag{50}$$

and so on. The solution in a series form is then given by

$$u(x) = 2 + 4\lambda + 16\lambda^2 + 80\lambda^3 + \cdots. \tag{51}$$

It is obvious from (51) that only one solution has been obtained by using the decomposition method. We recall that two answers have been obtained earlier by using the direct computation method given by

$$u(x) = \frac{1 \pm \sqrt{1 - 8\lambda}}{2\lambda}, \tag{52}$$

which, by using the binomial theorem to expand the square root , gives

$$u(x) = \frac{1 \pm (1 - 4\lambda - 8\lambda^2 - 32\lambda^3 - 160\lambda^4 + \cdots)}{2\lambda}, \tag{53}$$

so that $u(x)$ has the two expansions

$$u(x) = \frac{1}{\lambda} - \left(2 + 4\lambda + 16\lambda^2 + 80\lambda^3 + \cdots\right), \tag{54}$$

and

$$u(x) = \left(2 + 4\lambda + 16\lambda^2 + 80\lambda^3 + \cdots\right). \tag{55}$$

We can easily observe that the solution (51) obtained by using the decomposition method is consistent with the second solution (55) obtained by using the direct computation method. However, the decomposition method did not address any procedure to find the second solution (54).

Example 2. We next consider the nonlinear Fredholm integral equation

$$u(x) = \frac{7}{8}x + \frac{1}{2}\int_0^1 xtu^2(t)\,dt. \tag{56}$$

In this example the Adomian polynomials for the nonlinear term

$$F(u) = u^2(x), \tag{57}$$

are given by

$$\begin{cases} A_0 = u_0^2, \\ A_1 = 2u_0u_1 \\ A_2 = 2u_0u_2 + u_1^2 \\ A_3 = 2u_0u_3 + 2u_1u_2 \\ \vdots \end{cases} \tag{58}$$

Using the recursive algorithm (42) we find

$$u_0(x) = \frac{7}{8}x, \tag{59}$$

$$\begin{aligned} u_1(x) &= \frac{1}{2}x\int_0^1 tA_0(t)dt, \\ &= \frac{49}{512}x, \end{aligned} \tag{60}$$

$$\begin{aligned} u_2(x) &= \frac{1}{2}x\int_0^1 tA_1(t)dt, \\ &= \frac{343}{16384}x, \end{aligned} \tag{61}$$

and so on. The solution in a series form is given by

$$\begin{aligned} u(x) &= \frac{7}{8}x + \frac{49}{512}x + \frac{343}{16384}x + \cdots, \\ &= 0.875x + 0.0957031x + 0.02935x + \cdots, \\ &\simeq x. \end{aligned} \tag{62}$$

This is an example where the exact solution is not obtainable, hence we used few terms of the series to approximate the solution. We remark that two solutions $u(x) = x$ and $u(x) = 7x$ were obtained in (27) by using the direct computation method.

Example 3. As a third example we consider the nonlinear Fredholm integral equation

$$u(x) = 1 - \frac{1}{3}x + \int_0^1 xt^2u^3(t)\,dt. \tag{63}$$

The Adomian polynomials for the nonlinear term

$$F(u) = u^3(x), \tag{64}$$

have been determined before in (35). In this example we will use the *modified decomposition method.* Setting

$$u_0(x) \quad = \quad 1 \tag{65}$$

leads to

$$\begin{aligned} u_1(x) &= -\frac{1}{3}x + x\int_0^1 t^2A_0(t)dt \\ &= 0. \end{aligned} \tag{66}$$

As a result of (66) we obtain

$$u_k(x) = 0, \quad k \geq 1. \tag{67}$$

This yields the exact solution

$$u(x) = 1. \tag{68}$$

It is useful to note that the direct computation method, if used to solve this example, produces an algebraic equation of the third degree which may not be easy to solve. However, solving such an algebraic equation will give other solutions of the same equation that cannot be determined by Adomian decomposition method. In other words, the direct computation method produces all possible solutions, if exist, whereas the decomposition method is easier to use.

Exercises 6.2.2

Use the *decomposition method* or the *modified decomposition method* to solve the following nonlinear Fredholm integral equations:

1. $u(x) = 1 + \lambda \int_0^1 tu^2(t)\,dt, \quad \lambda \leq \frac{1}{2}$.

2. $u(x) = 1 + \lambda \int_0^1 t^3u^2(t)\,dt, \quad \lambda \leq 1.$

3. $u(x) = 2\sin x - \frac{\pi}{8} + \frac{1}{8}\int_0^{\pi/2} u^2(t)\,dt$.

4. $u(x) = 2\cos x - \frac{\pi}{8} + \frac{1}{8}\int_0^{\pi/2} u^2(t)\,dt.$

5. $u(x) = \sec x - x + x\int_0^{\pi/4} u^2(t)\,dt.$

6. $u(x) = \frac{3}{2}x + \frac{3}{8}\int_0^1 xu^2(t)dt.$

7. $u(x) = x^2 - \frac{1}{12} + \frac{1}{2}\int_0^1 tu^2(t)dt.$

8. $u(x) = x - \frac{\pi}{8} + \frac{1}{2}\int_0^1 \frac{1}{1+u^2(t)}dt.$

9. $u(x) = x - 1 + \frac{2}{\pi}\int_{-1}^1 \frac{1}{1+u^2(t)}dt.$

10. $u(x) = x - \frac{1}{4}\ln 2 + \frac{1}{2}\int_0^1 \frac{t}{1+u^2(t)}dt.$

11. $u(x) = \sin x + \cos x - \frac{\pi+2}{8} + \frac{1}{4}\int_0^{\pi/2} u^2(t)\,dt.$

12. $u(x) = \sinh x - 1 + \int_0^1 \left(\cosh^2(t) - u^2(t)\right)\,dt.$

13. $u(x) = \cos x + 2 - \int_0^1 \left(1 + \sin^2(t) + u^2(t)\right)\,dt.$

14. $u(x) = \sec x - x + \int_0^1 x\left(u^2(t) - \tan^2(t)\right)\,dt.$

6.3 Nonlinear Volterra Integral Equations

In this section we shall focus our study on simple cases of nonlinear Volterra integral equations, where other cases will be left for further studies. Based on our discussions in Chapters 3 and 5 we concluded that the *series solution method* and the *decomposition method* proved to be reliable techniques in handling successfully the linear Volterra integral equations and the Volterra integro-differential equations. It seems reasonable to use these methods in our study of the nonlinear Volterra integral equations. It is worth noting that both methods introduce one solution only and other solutions, if exist, are not handled by these methods.

It is to be noted that the decomposition method approaches the nonlinear Volterra equations generally by using the so called Adomian polynomials, that were introduced earlier in this chapter, hence we will skip details. The so called Adomian polynomials will be used to represent the involved nonlinear function $u^n(t), n \geq 2$ in a similar manner as discussed before. Although we calculate a finite number of the series terms in the decomposition method, but Adomian series was formally proved to be rapidly convergent.

In the following we will discuss the steps needed to use the series solution method and the decomposition method effectively.

6.3.1 The Series Solution Method

The nonlinear Volterra integral equations of the form

$$u(x) = f(x) + \lambda \int_0^x K(x,t)\, u^n(t)dt, \tag{69}$$

where the kernel $K(x,t)$ will be assumed a separable kernel, will be examined by using the *series solution method.* To use this method we should assume that $u(x)$ is analytic, hence it admits the Taylor expansion about $x = 0$ given by

$$u(x) = \sum_{n=0}^{\infty} a_n x^n. \tag{70}$$

Substituting (70) into both sides of (69), assuming that the kernel $K(x,t) = g(x)h(t)$, yields

$$\sum_{n=0}^{\infty} a_n x^n = f(x) + \lambda g(x) \int_0^x h(t) \left(\sum_{n=0}^{\infty} a_n t^n\right)^n dt, \tag{71}$$

or simply

$$a_0 + a_1 x + a_2 x^2 + \cdots = f(x) + \lambda g(x) \int_0^x h(t) \left(a_0 + a_1 t + a_2 t^2 + \cdots\right)^n dt, \tag{72}$$

such that the integral in (69) that includes the unknown function $u^n(x)$ is reduced to an easily calculable integrals in (72). Using the Taylor expansions for $f(x)$ and $g(x)$, integrating the resulting easy integrals at the right hand side of (72), and then equating the coefficients of like powers of x in both sides lead to a complete determination of the coefficients $a_0, a_1, a_2, \cdots$ of (70). Consequently, the solution $u(x)$ is readily obtained upon using (70).

As discussed before, the exact solution may be obtained if the resulting series is an expansion of a well known function, otherwise we use few terms of the series to achieve a numerical approximation of the solution for computational purposes. The method discussed above will be illustrated by discussing the following examples.

Example 1. We first consider the nonlinear Volterra integral equation

$$u(x) = x - \frac{1}{4}x^4 + \int_0^x t u^2(t)\, dt. \tag{73}$$

Substituting $u(x)$ in a series form given by (70) into both sides of (73) yields

$$\begin{aligned} a_0 &+ a_1 x + a_2 x^2 + a_3 x^3 + \cdots \\ &= x - \frac{1}{4}x^4 + \int_0^x t\left[a_0 + a_1 t + a_2 t^2 + a_3 t^3 + \cdots\right]^2 dt, \end{aligned} \tag{74}$$

or equivalently

$$\begin{aligned} a_0 &+ a_1 x + a_2 x^2 + a_3 x^3 + \cdots \\ &= x - \frac{1}{4}x^4 + \int_0^x t\left[a_0^2 + 2a_0 a_1 t + \left(2a_0 a_2 + a_1^2\right) t^2 + \cdots\right] dt. \end{aligned} \tag{75}$$

Integrating the integral at the right hand side of (75) we find

$$\begin{aligned} a_0 \quad &+ \quad a_1 x + a_2 x^2 + a_3 x^3 + \cdots \\ &= \quad x - \frac{1}{4}x^4 + \frac{1}{2}a_0^2 x^2 + \frac{2}{3}a_0 a_1 x^3 + \frac{1}{4}\left(a_1^2 + 2a_0 a_2\right)x^4 + \cdots . \end{aligned} \tag{76}$$

Equating the coefficients of like powers of x in both sides yields

$$a_1 = 1, \quad a_n = 0, \quad \text{for} \quad n \neq 1 \tag{77}$$

Consequently, the exact solution is given by

$$u(x) = x, \tag{78}$$

upon substituting (77) into (70). This can be justified through substitution into Eq. (73).

Example 2. We next consider the nonlinear Volterra integral equation

$$u(x) = e^x - \frac{1}{3}xe^{3x} + \frac{1}{3}x + \int_0^x xu^3(t)dt. \tag{79}$$

Substituting the series (70) into both sides of (79) noting that

$$u^3(x) = a_0^3 + 3a_0^2 a_1 x + (3a_0 a_1^2 + 3a_0^2 a_2)x^2 + \cdots \tag{80}$$

gives

$$\begin{aligned} a_0 \quad +a_1 x + \quad & a_2 x^2 + a_3 x^3 + \cdots = \left(1 + x + \frac{x^2}{2!} + \frac{x^3}{3!} + \cdots\right) \\ - \quad & \frac{1}{3}x\left(1 + 3x + \frac{9}{2}x^2 + \frac{27}{3!}x^3 + \cdots\right) + \frac{1}{3}x \\ + \quad & x\int_0^x \left[{a_0}^3 + 3{a_0}^2 a_1 t + (3a_0 {a_1}^2 + 3{a_0}^2 a_2)t^2 + \cdots\right] dt. \end{aligned} \tag{81}$$

It is obvious that the nonlinear Volterra integral equation given in this example has been transformed to easily calculable integrals involving terms of the form t^n, where $n \geq 0$. Besides, the Taylor series of the functions e^x and e^{3x} have been used. Evaluating the integral at the right hand side of (81), collecting like terms and equating the coefficients of like powers of x we find

$$\begin{aligned} a_0 &= 1 \\ a_1 &= 1 \\ a_2 &= \frac{1}{2!} \\ a_3 &= \frac{1}{3!} \\ &\vdots \\ a_n &= \frac{1}{n!} \quad \text{for } n \geq 0 \end{aligned} \tag{82}$$

This gives the solution in a series form

$$u(x) = 1 + x + \frac{1}{2!}x^2 + \frac{1}{3!}x^3 + \cdots \tag{83}$$

so that

$$u(x) = e^x, \tag{84}$$

is the exact solution

Exercises 6.3.1

Use the *series solution method* to solve the following nonlinear Volterra integral equations .

1. $u(x) = x^2 + \frac{1}{10}x^5 - \frac{1}{2}\int_0^x u^2(t)\,dt,$
2. $u(x) = x^2 + \frac{1}{12}x^6 - \frac{1}{2}\int_0^x tu^2(t)\,dt,$
3. $u(x) = 1 - x^2 - \frac{1}{3}x^3 + \int_0^x u^2(t)\,dt,$
4. $u(x) = 1 - x + x^2 - \frac{2}{3}x^3 - \frac{1}{5}x^5 + \int_0^x u^2(t)\,dt,$
5. $u(x) = x^2 + \frac{1}{14}x^7 - \frac{1}{2}\int_0^x u^3(t)\,dt,$

6. $u(x) = \frac{1}{2} + e^{-x} - \frac{1}{2}e^{-2x} - \int_0^x u^2(t)\,dt,$

7. $u(x) = 1 - \frac{3}{2}x^2 - x^3 - \frac{1}{4}x^4 + \int_0^x u^3(t)\,dt,$

8. $u(x) = \sin x - \frac{1}{2}x + \frac{1}{4}\sin 2x + \int_0^x u^2(t)\,dt,$

9. $u(x) = \cos x - \frac{1}{2}x - \frac{1}{4}\sin 2x + \int_0^x u^2(t)\,dt,$

10. $u(x) = e^x + \frac{1}{2}x(e^{2x} - 1) - \int_0^x xu^2(t)\,dt,$

6.3.2 The Decomposition Method

The purpose of this brief section is to describe how the Adomian decomposition method can be applied to nonlinear Volterra integral equations. Even though the method does not discuss the existence and the uniqueness concepts, but it provides a reliable and powerful technique to handle nonlinear equations. As discussed before, the method allows us to obtain the exact solution as an infinite series of functions.

The method has been introduced in details in solving nonlinear Fredholm integral equations. For this reason, we will outline a brief framework of the method that will be implemented for nonlinear Volterra integral equations. Recall that we will focus our study on the nonlinear Volterra integral equations of the form

$$u(x) = f(x) + \lambda \int_0^x K(x,t)\,u^n(t)dt, \tag{85}$$

where the kernel is assumed a separable kernel. We usually represent the solution $u(t)$ of (85) by the series

$$u(x) = \sum_{n=0}^{\infty} u_n(x), \tag{86}$$

and the nonlinear term $u^n(t)$, under the integral sign of the equation (85), by the polynomial series

$$u^n(t) = \sum_{n=0}^{\infty} A_n(t), \tag{87}$$

where the $A_n(t)\, n \geq 0$ are the so called Adomian polynomials. The Adomian polynomials can be established by using the schemes

$$\begin{cases} A_0 &= F(u_0), \\ A_1 &= u_1\, F'(u_0), \\ A_2 &= u_2\, F'(u_0) + \frac{u_1^2}{2!}\, F''(u_0), \\ A_3 &= u_3\, F'(u_0) + u_1 u_2 F''(u_0) + \frac{u_1^3}{3!}\, F'''(u_0), \\ \cdot \\ \cdot \\ \cdot \end{cases} \tag{88}$$

We introduced in the previous section several examples explaining how we can generate Adomian polynomials .

Substituting (86) and (87) into (85) yields

$$\sum_{n=0}^{\infty} u_n(x) = f(x) + \lambda \int_0^x K(x,t) \left(\sum_{n=0}^{\infty} A_n(t) \right) dt. \tag{89}$$

or simply

$$\begin{aligned} u_0(x) &+ u_1(x) + u_2(x) + \cdots \\ &= f(x) + \lambda \int_0^x K(x,t) \left[A_0(t) + A_1(t) + A_2(t) + \cdots\right] dt. \end{aligned} \tag{90}$$

The components $u_0(x), u_1(x), u_2(x), \cdots$ are completely determined by using the recurrent scheme

$$\begin{cases} u_0(x) &= f(x), \\ u_1(x) &= \lambda \int_0^x K(x,t) A_0(t) dt, \\ u_2(x) &= \lambda \int_0^x K(x,t) A_1(t) dt, \\ \vdots \\ u_{n+1}(x) = \lambda \int_a^b K(x,t) A_n(t) dt, & n \geq 0. \end{cases} \tag{91}$$

Consequently the solution of (85) in a series form is immediately determined by using (86). As indicated earlier, the series obtained may yield the exact solution in a closed form, or a truncated series $\sum_{n=1}^{k} u_n(x)$ may be used if a numerical approximation is desired.

It is worth noting that the decomposition method encounters computational difficulties if the nonhomogeneous part $f(x)$ is not a polynomial of few terms. Other cases where $f(x)$, is not a polynomial, the *modified decomposition method* plays a major role in minimizing the size of calculations.

In the following examples, we will illustrate the decomposition technique and the modified decomposition method for handling the nonlinear Volterra integral equations.

Example 1. We first consider the nonlinear Volterra integral equation

$$u(x) = x + \frac{1}{5}x^5 - \int_0^x tu^3(t)\,dt. \tag{92}$$

We start by setting the zeroth component

$$u_0(x) = x + \frac{1}{5}x^5, \tag{93}$$

so that the first component is obtained by

$$u_1(x) = -\int_0^x tA_0(t)\,dt, \tag{94}$$

which gives

$$u_1(x) = -\frac{1}{5}x^5 - \frac{1}{15}x^9 - \frac{3}{325}x^{13} - \frac{1}{2125}x^{17}, \tag{95}$$

upon using

$$A_0(t) = (t + \frac{1}{5}t^5)^3. \tag{96}$$

It can be easily observed that by cancelling the noise terms $\frac{1}{5}x^5$ and $-\frac{1}{5}x^5$ between $u_0(x)$ and $u_1(x)$, and justifying that the remaining term of $u_0(x)$ satisfies the equation, lead to the exact solution

$$u(x) = x. \tag{97}$$

We point out here that the exact solution can also be obtained by using the modified decomposition method simply by setting $u_0(x) = x$.

Example 2. We next consider the nonlinear Volterra integral equation

$$u(x) = 2x - \frac{1}{12}x^4 + \frac{1}{4}\int_0^x (x-t)u^2(t)\, dt. \tag{98}$$

To minimize the calculations volume, we will use the modified decomposition method in this example. For this reason we split $f(x)$ between the two components $u_0(x)$ and $u_1(x)$, hence we set

$$u_0(x) = 2x. \tag{99}$$

Consequently, the first component is defined by

$$u_1(x) = -\frac{1}{12}x^4 + \frac{1}{4}\int_0^x (x-t)A_0(t)\, dt, \tag{100}$$

or simply by

$$u_1(x) = 0, \tag{101}$$

upon using

$$A_0(t) = (2t)^2. \tag{102}$$

This defines the other components by

$$u_k(x) = 0, \quad \text{for } k \geq 1. \tag{103}$$

The exact solution

$$u(x) = 2x, \tag{104}$$

follows immediately.

Example 3. It seems reasonable to compare the *series solution method* and the *Adomian decomposition method* by solving Example 2 in the previous subsection on page 175 given by

$$u(x) = e^x - \frac{1}{3}xe^{3x} + \frac{1}{3}x + \int_0^x xu^3(t)dt. \tag{105}$$

Applying the standard decomposition method will result in a considerable amount of difficulties in integrating and forming the Adomian polynomials. It is useful to consider using the *modified decomposition method.* Splitting $f(x)$ between the first two components yields

$$u_0(x) = e^x, \tag{106}$$

and

$$u_1(x) = -\frac{1}{3}xe^{3x} + \frac{1}{3}x + \int_0^x xA_0(t)dt, \tag{107}$$

or equivalently

$$u_1(x) = 0 \tag{108}$$

upon using

$$A_0(t) = (e^{3t}). \tag{109}$$

This defines the other components by

$$u_k(x) = 0, \quad \text{for } k \geq 1. \tag{110}$$

Accordingly, the exact solution

$$u(x) = e^x, \tag{111}$$

is readily obtained. It is clear that the modified decomposition method deals with the nonlinearity in an efficient way.

Exercises 6.3.2

Use the *decomposition method* or the *modified decomposition method* to solve the following nonlinear Volterra integral equations by finding the exact solution or few terms of the series solution

1. $u(x) = 3x + \frac{1}{24}x^4 - \frac{1}{18}\int_0^x (x-t)u^2(t)\,dt.$
2. $u(x) = 2x - \frac{1}{2}x^4 + \frac{1}{4}\int_0^x u^3(t)\,dt.$
3. $u(x) = \sin x + \frac{1}{8}\sin(2x) - \frac{1}{4}x + \frac{1}{2}\int_0^x u^2(t)\,dt\ .$
4. $u(x) = x^2 + \frac{1}{5}x^5 - \int_0^x u^2(t)\,dt.$

5. $u(x) = x + \int_0^x (x-t)u^3(t)\,dt.$

6. $u(x) = 1 + \int_0^x (x-t)^2 u^2(t)\,dt.$

7. $u(x) = 1 + \int_0^x (x-t)^2 u^3(t)\,dt.$

8. $u(x) = x + \int_0^x (x-t)^2 u^2(t)\,dt.$

9. $u(x) = 1 + \int_0^x \left(t + u^2(t)\right)\,dt.$

10. $u(x) = 1 + \int_0^x \left(t^2 + u^2(t)\right)\,dt.$

11. $u(x) = \sec x + \tan x + x - \int_0^x \left(1 + u^2(t)\right)\,dt,\ x < \frac{\pi}{2}.$

12. $u(x) = \tan x - \frac{1}{4}\sin(2x) - \frac{x}{2} + \int_0^x \frac{1}{1+u^2(t)}\,dt,\ x < \frac{\pi}{2}.$

Appendix A

Table of Integrals

A.1 Basic Forms

1. $\int x^n \, dx = \frac{1}{n+1} x^{n+1} + C, n \neq -1.$

2. $\int \frac{1}{x} \, dx = \ln|x| + C.$

3. $\int e^x \, dx = e^x + C.$

4. $\int e^{-x} \, dx = -e^{-x} + C.$

5. $\int \frac{1}{1+x^2} \, dx = \tan^{-1} x + C.$

6. $\int \frac{1}{\sqrt{1-x^2}} \, dx = \sin^{-1} x + C.$

7. $\int \cos x \, dx = \sin x + C.$

8. $\int \sin x \, dx = -\cos x + C.$

9. $\int \tan x \, dx = -\ln|\cos x| + C.$

10. $\int \cot x \, dx = \ln|\sin x| + C.$

11. $\int \tan x \sec x \, dx = \sec x + C.$

12. $\int \sec^2 x \, dx = \tan x + C.$

13. $\int \csc^2 x \, dx = -\cot x + C.$

A.2 Trigonometric Forms

1. $\int \sin^2 x \, dx = \frac{1}{2}x - \frac{1}{4}\sin 2x + C.$

2. $\int \cos^2 x \, dx = \frac{1}{2}x + \frac{1}{4}\sin 2x + C.$

3. $\int \sin^3 x \, dx = -\frac{1}{3}\cos x \left(2 + \sin^2 x\right) + C.$

4. $\int \cos^3 x \, dx = \frac{1}{3}\sin x \left(2 + \cos^2 x\right) + C.$

5. $\int \tan^2 x \, dx = \tan x - x + C.$

6. $\int \cot^2 x \, dx = -\cot x - x + C.$

7. $\int \tan^3 x \, dx = \frac{1}{2}\tan^2 x + \ln|\cos x| + C.$

8. $\int \cot^3 x \, dx = -\frac{1}{2}\cot^2 x - \ln|\sin x| + C.$

9. $\int x \sin x \, dx = \sin x - x\cos x + C.$

10. $\int x \cos x \, dx = \cos x + x \sin x + C.$

11. $\int x^2 \sin x \, dx = 2x \sin x - \left(x^2 - 2\right)\cos x + C.$

12. $\int x^2 \cos x \, dx = 2x \cos x + \left(x^2 - 2\right)\sin x + C.$

A.3 Inverse Trigonometric Forms

1. $\int \sin^{-1}x\, dx = x\sin^{-1}x + \sqrt{1-x^2} + C.$

2. $\int \cos^{-1}x\, dx = x\cos^{-1}x - \sqrt{1-x^2} + C.$

3. $\int \tan^{-1}x\, dx = x\tan^{-1}x - \frac{1}{2}\ln(1+x^2) + C.$

4. $\int x\sin^{-1}x\, dx = \frac{1}{4}[(2x^2-1)\sin^{-1}x + x\sqrt{1-x^2}] + C.$

5. $\int x\cos^{-1}x\, dx = \frac{1}{4}[(2x^2-1)\cos^{-1}x - x\sqrt{1-x^2}] + C.$

6. $\int x\tan^{-1}x\, dx = \frac{1}{2}[(x^2+1)\tan^{-1}x - x] + C.$

7. $\int \sec^{-1}x\, dx = x\sec^{-1}x - \ln(x + \sqrt{x^2-1}) + C.$

8. $\int x\sec^{-1}x\, dx = \frac{1}{2}[x^2\sec^{-1}x - \sqrt{x^2-1}] + C.$

A.4 Exponential and Logarithmic Forms

1. $\int e^{ax}\, dx = \frac{1}{a}e^{ax} + C.$

2. $\int xe^{ax}\, dx = \frac{1}{a^2}(ax-1)e^{ax} + C.$

3. $\int x^2e^{ax}\, dx = \frac{1}{a^3}(a^2x^2 - 2ax + 2)e^{ax} + C.$

4. $\int x^3e^{ax}\, dx = \frac{1}{a^4}(a^3x^3 - 3a^2x^2 + 6ax - 6)e^{ax} + C.$

5. $\int e^x \sin x\, dx = \frac{1}{2}(\sin x - \cos x)e^x + C.$

6. $\int e^x \cos x\, dx = \frac{1}{2}(\sin x + \cos x)e^x + C.$

7. $\int \ln x \, dx = x \ln x - x + C.$

8. $\int x \ln x \, dx = \frac{1}{2}x^2(\ln x - \frac{1}{2}) + C.$

9. $\int x^2 \ln x \, dx = \frac{1}{3}x^3(\ln x - \frac{1}{3}) + C.$

10. $\int \frac{1}{x \ln x} \, dx = \ln |\ln x| + C.$

A.5 Hyperbolic Functions Forms

1. $\int \sinh x \, dx = \cosh x + C.$

2. $\int \cosh x \, dx = \sinh x + C.$

3. $\int x \sinh x \, dx = x \cosh x - \sinh x + C.$

4. $\int x \cosh x \, dx = x \sinh x - \cosh x + C.$

5. $\int \sinh^2 x \, dx = \frac{1}{2}(\sinh x \cosh x - x) + C.$

6. $\int \cosh^2 x \, dx = \frac{1}{2}(\sinh x \cosh x + x) + C.$

7. $\int \operatorname{sech}^2 x \, dx = \tanh x + C.$

8. $\int \tanh x \, dx = \ln \cosh x + C.$

9. $\int \coth x \, dx = \ln \sinh x + C.$

10. $\int \operatorname{sech} x \tanh x \, dx = -\operatorname{sech} x + C.$

Appendix B

Integrals of Irrational Functions

B.1 First Type of Forms

Integrals Involving $\dfrac{t^n}{\sqrt{x-t}}$, n is an integer, $n \geq 0$

1. $\displaystyle\int_0^x \frac{1}{\sqrt{x-t}}\,dt = 2\sqrt{x}.$

2. $\displaystyle\int_0^x \frac{t}{\sqrt{x-t}}\,dt = \frac{4}{3}x^{\frac{3}{2}}.$

3. $\displaystyle\int_0^x \frac{t^2}{\sqrt{x-t}}\,dt = \frac{16}{15}x^{\frac{5}{2}}.$

4. $\displaystyle\int_0^x \frac{t^3}{\sqrt{x-t}}\,dt = \frac{32}{35}x^{\frac{7}{2}}.$

5. $\displaystyle\int_0^x \frac{t^4}{\sqrt{x-t}}\,dt = \frac{256}{315}x^{\frac{9}{2}}.$

6. $$\int_0^x \frac{t^5}{\sqrt{x-t}}\,dt = \frac{512}{693}x^{\frac{11}{2}}.$$

7. $$\int_0^x \frac{t^6}{\sqrt{x-t}}\,dt = \frac{2048}{3003}x^{\frac{13}{2}}.$$

8. $$\int_0^x \frac{t^7}{\sqrt{x-t}}\,dt = \frac{4096}{6435}x^{\frac{15}{2}}.$$

9. $$\int_0^x \frac{t^8}{\sqrt{x-t}}\,dt = \frac{65536}{109395}x^{\frac{17}{2}}.$$

B.2 Second Type of Forms

Integrals Involving $\frac{t^{\frac{n}{2}}}{\sqrt{x-t}}$, n is an odd integer, $n \geq 1$

1. $$\int_0^x \frac{t^{\frac{1}{2}}}{\sqrt{x-t}}\,dt = \frac{1}{2}\pi x.$$

2. $$\int_0^x \frac{t^{\frac{3}{2}}}{\sqrt{x-t}}\,dt = \frac{3}{8}\pi x^2.$$

3. $$\int_0^x \frac{t^{\frac{5}{2}}}{\sqrt{x-t}}\,dt = \frac{5}{16}\pi x^3.$$

4. $$\int_0^x \frac{t^{\frac{7}{2}}}{\sqrt{x-t}}\,dt = \frac{35}{128}\pi x^4.$$

5. $$\int_0^x \frac{t^{\frac{9}{2}}}{\sqrt{x-t}}\,dt = \frac{63}{256}\pi x^5.$$

6. $$\int_0^x \frac{t^{\frac{11}{2}}}{\sqrt{x-t}}\,dt = \frac{231}{1024}\pi x^6.$$

Appendix C

Series Representations

C.1 Exponential Functions Series

1. $e^x = 1 + x + \frac{x^2}{2!} + \frac{x^3}{3!} + \frac{x^4}{4!} + \cdots.$
2. $e^{-x} = 1 - x + \frac{x^2}{2!} - \frac{x^3}{3!} + \frac{x^4}{4!} - \cdots.$
3. $e^{-x^2} = 1 - \frac{x^2}{1!} + \frac{x^4}{2!} - \frac{x^6}{3!} + \cdots.$
4. $a^x = 1 + x\ln a + \frac{1}{2!}(x\ln a)^2 + \frac{1}{3!}(x\ln a)^3 + \cdots, a > 0.$

C.2 Trigonometric Functions Series

1. $\sin x = x - \frac{x^3}{3!} + \frac{x^5}{5!} - \frac{x^7}{7!} + \cdots.$
2. $\cos x = 1 - \frac{x^2}{2!} + \frac{x^4}{4!} - \frac{x^6}{6!} + \cdots.$
3. $\tan x = x + \frac{x^3}{3} + \frac{2x^5}{15} + \frac{17x^7}{315} + \cdots.$

4. $\sin x + \cos x = (1+x) - \left(\frac{x^2}{2!} + \frac{x^3}{3!}\right) + \left(\frac{x^4}{4!} + \frac{x^5}{5!}\right) - \cdots.$

C.3 Hyperbolic Functions Series:

1. $\sinh x = x + \frac{x^3}{3!} + \frac{x^5}{5!} + \frac{x^7}{7!} + \cdots.$
2. $\cosh x = 1 + \frac{x^2}{2!} + \frac{x^4}{4!} + \frac{x^6}{6!} + \cdots.$
3. $\tanh x = x - \frac{x^3}{3} + \frac{2x^5}{15} - \frac{17x^7}{315} + \cdots.$

C.4 Logarithmic Functions Series

1. $\ln x = (x-1) - \frac{1}{2}(x-1)^2 + \frac{1}{3}(x-1)^3 - \frac{1}{4}(x-1)^4 + \cdots, 0 < x \leq 2.$
2. $\ln(1+x) = x - \frac{1}{2}x^2 + \frac{1}{3}x^3 - \frac{1}{4}x^4 + \cdots, \quad -1 < x \leq 1.$
3. $\ln(1-x) = -\left(x + \frac{1}{2}x^2 + \frac{1}{3}x^3 + \frac{1}{4}x^4 + \cdots\right), \quad -1 \leq x < 1.$

Appendix D

The Error and Gamma Functions

D.1 The Error Function

The *error function* **erf(x) is defined by:**

1. $erf(x) = \frac{2}{\sqrt{\pi}} \int_0^x e^{-u^2} du.$

2. $erf(x) = \frac{2}{\sqrt{\pi}} \left(x - \frac{x^3}{3} + \frac{x^5}{5 \cdot 2!} - \frac{x^7}{7 \cdot 3!} + \cdots \right).$

D.2 The Complementary Error Function

erfc(x) is defined by:

1. $erfc(x) = \frac{2}{\sqrt{\pi}} \int_x^\infty e^{-u^2} du.$

2. $erf(x) + erfc(x) = 1.$

3. $erfc(x) = 1 - \frac{2}{\sqrt{\pi}} \left(x - \frac{x^3}{3} + \frac{x^5}{5 \cdot 2!} - \frac{x^7}{7 \cdot 3!} + \cdots \right).$

4. $erf(-x) = -erf(x)$.

5. $erf(0) = 0$.

6. $erf(\infty) = 1$.

D.3 The Gamma Function $\Gamma(x)$

1. $\Gamma(x) = \int_0^\infty t^{x-1} e^{-t} dt$.

2. $\Gamma(x+1) = x\Gamma(x)$.

3. $\Gamma(1) = 1, \Gamma(n+1) = n!$, **n is an integer.**

4. $\Gamma(x)\Gamma(1-x) = \frac{\pi}{\sin \pi x}$.

5. $\Gamma(\frac{1}{2}) = \sqrt{\pi}$.

6. $\Gamma(\frac{3}{2}) = \frac{1}{2}\sqrt{\pi}$.

7. $\Gamma(\frac{1}{2})\Gamma(-\frac{1}{2}) = -2\pi$.

Answers To Exercises

Exercises 1.2

1. Fredholm, linear, nonhomogeneous
2. Volterra, linear, nonhomogeneous
3. Volterra, nonlinear, nonhomogeneous
4. Fredholm, linear, homogeneous
5. Fredholm, linear, nonhomogeneous
6. Fredholm, nonlinear, nonhomogeneous
7. Fredholm, nonlinear, nonhomogeneous
8. Fredholm, linear, nonhomogeneous
9. Volterra, nonlinear, nonhomogeneous
10. Volterra, linear, nonhomogeneous
11. Volterra integro-differential equation, nonlinear
12. Fredholm integro-differential equation, linear
13. Volterra integro-differential equation, nonlinear
14. Fredholm integro-differential equation, linear
15. Volterra integro-differential equation, linear
16. $u(x) = 1 + \int_0^x 4u(t)dt$
17. $u(x) = 1 + \int_0^x 3t^2 u(t)dt$
18. $u(x) = 4 + \int_0^x u^2(t)dt$
19. $u'(x) = 1 + \int_0^x 4tu^2(t)dt, \; u(0) = 2$
20. $u'(x) = 1 + \int_0^x 2tu(t)dt, \; u(0) = 0$

Exercises 1.4

1. $\int_0^x 3(x-t)^2 u(t)dt$

2. $2xe^{x^3} - e^{x^2} + \int_x^{x^2} te^{xt}dt$

3. $\int_0^x 4(x-t)^3 u(t)dt$

4. $4\sin 5x - \sin 2x + \int_x^{4x} \cos(x+t)dt$

5. $u'''(x) = 2u(x),\ u(0) = u'(0) = 1,\ u''(0) = 0$

6. $u''(x) + u(x) = e^x,\ u(0) = u'(0) = 1$

7. $u''(x) - u(x) = 0,\ u(0) = 0,\ u'(0) = 1$

8. $u''(x) - u(x) = \cos x,\ u(0) = -1,\ u'(0) = 1$

9. $u''(x) - u'(x) - 2u(x) = 10,\ u(0) = 2,\ u'(0) = 5$

10. $u''(x) - 5u'(x) + 6u(x) = 0,\ u(0) = -5,\ u'(0) = -19$

11. $u'(x) + u(x) = \sec^2 x,\ u(0) = 0$

12. $u'''(x) - 3u''(x) - 6u'(x) + 5u(x) = 0,$

    $u(0) = 1, u'(0) = 4, u''(0) = 23$

13. $u'''(x) - 4u(x) = 24x,$

    $u(0) = u'(0) = 0, u''(0) = 2$

14. $u^{iv}(x) - u(x) = 0,$

    $u(0) = u'(0) = 0, u''(0) = 2, u'''(0) = 0$

Exercises 1.5

1. $u(x) = -1 - \int_0^x u(t)dt, \text{ where } y'(x) = u(x)$

2. $u(x) = x + \int_0^x u(t)dt, \text{ where } y'(x) = u(x)$

3. $u(x) = \sec^2(x) - \int_0^x u(t)dt, \text{ where } y'(x) = u(x)$

In problems **4** – **10**, set $y''(x) = u(x)$

4. $u(x) = -1 - \int_0^x (x-t)u(t)dt,$

5. $u(x) = 1 + x + \int_0^x (x-t)u(t)dt,$

6. $u(x) = -11 - 6x - \int_0^x [5 + 6(x-t)]\, u(t)dt$

7. $u(x) = -1 - \int_1^x u(t)dt$

8. $u(x) = -1 + 4x + \int_0^x [2(x-t) - 1]\, u(t)dt$

9. $u(x) = \sin x - \int_0^x (x-t)u(t)dt$

10. $u(x) = x - \sin x + xe^x - e^x - \int_0^x [(x-t)e^x - \sin x]\, u(t)dt$

In problems **11** – **15**, set $y'''(x) = u(x)$

11. $u(x) = 2x - x^2 + \int_0^x \left[1 + (x-t) - \frac{1}{2}(x-t)^2\right] u(t)dt$

12. $u(x) = -3x - 4\int_0^x (x-t)u(t)dt$

13. $u(x) = 2 + x - \frac{1}{2}x^2 - \frac{1}{6}x^3 - \int_0^x \left[2(x-t) + \frac{1}{6}(x-t)^3\right] u(t)dt$

14. $u(x) = 1 - \frac{1}{2}x^2 + \frac{1}{3!}\int_0^x (x-t)^3 u(t)dt$

15. $u(x) = 2e^x - 1 - x - \int_0^x (x-t)u(t)dt$

Exercises 1.6

1. $u(x) = \sin x + \int_0^1 K(x,t)u(t)\,dt,$

where the kernel $K(x,t)$ is defined by

$$K(x,t) = \begin{cases} 4t(1-x) & 0 \le t \le x \\ 4x(1-t) & x \le t \le 1 \end{cases}$$

2. $u(x) = 1 + \int_0^1 K(x,t)u(t)dt,$

where the kernel $K(x,t)$ is defined by

$$K(x,t) = \begin{cases} 2xt(1-x) & 0 \le t \le x \\ 2x^2(1-t) & x \le t \le 1 \end{cases}$$

3. $u(x) = (2x-1) + \int_0^1 K(x,t)u(t)dt,$

where the kernel $K(x,t)$ is defined by

$$K(x,t) = \begin{cases} t(1-x) & 0 \le t \le x \\ x(1-t) & x \le t \le 1 \end{cases}$$

4. $u(x) = (x-1) + \int_0^1 K(x,t)u(t)dt,$

where the kernel $K(x,t)$ is defined by

$$K(x,t) = \begin{cases} t & 0 \le t \le x \\ x & x \le t \le 1 \end{cases}$$

Exercises 2.2

1. $u(x) = 4x$ 2. $u(x) = x^3$

3. $u(x) = x^2 + \frac{3}{8}x$
4. $u(x) = 1 + e^x$
5. $u(x) = \sin x$
6. $u(x) = \cos x$
7. $u(x) = \cos(4x)$
8. $u(x) = \sinh x$
9. $u(x) = 2e^{2x}$
10. $u(x) = \sec^2 x$
11. $u(x) = \sin x$
12. $u(x) = \tan x$
13. $u(x) = \tan^{-1} x$
14. $u(x) = \cosh x$
15. $u(x) = \frac{1}{1+x^2}$
16. $u(x) = \frac{1}{\sqrt{1-x^2}}$
17. $u(x) = \frac{1}{1+x^2}$
18. $u(x) = \cos^{-1} x$
19. $u(x) = x\tan^{-1} x$
20. $u(x) = x\sin^{-1} x + 1$

Exercises 2.3

1. $u(x) = xe^x$
2. $u(x) = x^2 - 2x + 1$
3. $u(x) = x \sin x$
4. $u(x) = e^{2x}$
5. $u(x) = 1 + \sec^2 x$
6. $u(x) = \sin(2x)$
7. $u(x) = x^2 - \frac{5}{18}x - \frac{5}{36}$
8. $u(x) = \sin x + \cos x$
9. $u(x) = \sec x \tan x$
10. $u(x) = x^2$
11. $u(x) = \sin x$
12. $u(x) = 1 + \frac{1}{2} \ln x$
13. $u(x) = x^3$
14. $u(x) = 1 + \frac{\pi}{4} \sec^2 x$

Exercises 2.4

1. $u(x) = x$
2. $u(x) = x^3$
3. $u(x) = 4x$
4. $u(x) = 1 + 2x$
5. $u(x) = 2 \sin x$
6. $u(x) = \sec^2 x$
7. $u(x) = \sec x \tan x$
8. $u(x) = \cosh x$
9. $u(x) = e^x$
10. $u(x) = \sin x$

Exercises 2.5

1. $u(x) = 2x$
2. $u(x) = 1 - \frac{\pi}{10}\cos x$
3. $u(x) = x + 1$
4. $u(x) = \sin x + \cos x$
5. $u(x) = x^2$
6. $u(x) = x^3$
7. $u(x) = \sin x + \cos x$
8. $u(x) = 1 + \frac{\pi}{2}\sin x$
9. $u(x) = 1 + \frac{\pi}{4}\sec^2 x$
10. $u(x) = 1 + \frac{\pi}{12}\sec x \tan x$

Exercises 2.7

1. $u(x) = A$, A is a constant
2. $u(x) = 2Ax$
3. $u(x) = Ax$, A is an a constant
4. $u(x) = A\cos x$
5. $u_1(x) = \frac{2}{\pi}A\,(\sin x + \cos x)$ $\quad u_2(x) = \frac{2}{\pi}A\,(\sin x - \cos x)$
6. $u_1(x) = u_2(x) = \frac{2}{\pi}\,(A\sin x + B\cos x)$
7. $u(x) = A\sec x$
8. $u(x) = A\sec^2 x$
9. $u(x) = \frac{2}{\pi - 2}A\sin^{-1} x$
10. $u(x) = \alpha(3 - \frac{3}{2}x)$

Exercises 3.2

1. $u(x) = 4x$
2. $u(x) = 1 + 2x$
3. $u(x) = e^{-x}$
4. $u(x) = \sinh x$
5. $u(x) = \sin(3x)$
6. $u(x) = \cos(2x)$
7. $u(x) = \sin x + \cos x$
8. $u(x) = \cos x - \sin x$
9. $u(x) = e^x$
10. $u(x) = e^{-x}$
11. $u(x) = 2\cosh x$
12. $u(x) = 2e^x - 1$
13. $u(x) = 2\cos x - 1$
14. $u(x) = 2\cosh x - 1$
15. $u(x) = \cos x$
16. $u(x) = \sec^2 x$
17. $u(x) = \cosh x$
18. $u(x) = \sinh x$
19. $u(x) = x^3$
20. $u(x) = \sec x \tan x$

Exercises 3.3

1. $u(x) = 2x + 3x^2$
2. $u(x) = 1 + x + x^2$
3. $u(x) = \sin x + \cos x$
4. $u(x) = 1 + x$
5. $u(x) = -e^{-x}$
6. $u(x) = e^{-2x}$
7. $u(x) = e^{x}$
8. $u(x) = \sinh x$
9. $u(x) = 2\cos x - 1$
10. $u(x) = \cos x - \sin x$
11. $u(x) = \sinh x$
12. $u(x) = \sin x$

Exercises 3.4

1. $u(x) = e^{-3x}$
2. $u(x) = \cosh x$
3. $u(x) = \cos x - \sin x$
4. $u(x) = e^{x} - 1$
5. $u(x) = e^{x}$
6. $u(x) = \frac{1}{2}(\cos x + \cosh x)$
7. $u(x) = \frac{1}{2}(\sin x + \sinh x)$
8. $u(x) = 2\cosh x - 2$
9. $u(x) = x$
10. $u(x) = x - x^2$

Exercises 3.5

1. $u(x) = e^{-x}$
2. $u(x) = \cos(3x)$
3. $u(x) = e^{2x}$
4. $u(x) = e^{-\frac{1}{4}x}$
5. $u(x) = 2\cos x$
6. $u(x) = e^{-x^2}$
7. $u(x) = \sinh x$
8. $u(x) = \cos x$
9. $u(x) = \cos x + \sin x$
10. $u(x) = \cos x - \sin x$
11. $u(x) = 1 + e^{x}$
12. $u(x) = 1 - \sinh x$

Exercises 3.6

1. $u(x) = e^{x} - 1$
2. $u(x) = e^{x} - x - 1$
3. $u(x) = x - \sin x$
4. $u(x) = -x + \sinh x$
5. $u(x) = -1 + \cos x$
6. $u(x) = 1 - x$

7. $u(x) = e^{2x}$ 8. $u(x) = 2 + e^x$

9. $u(x) = 1 + \cos x$ 10. $u(x) = 1 - \sin x$

11. $u(x) = 1 + \cosh x$ 12. $u(x) = -1 + \cosh x$

Exercises 3.8

1. $u(x) = 2x$ 2. $u(x) = e^{-x}$

3. $u(x) = e^x$ 4. $u(x) = \sinh x$

5. $u(x) = \cos x$ 6. $u(x) = \sec^2 x$

Exercises 4.2.1

1. $u(x) = \frac{1}{6}(1 + x)$ 2. $u(x) = \frac{1}{6} - \frac{1}{63}x^2$

3. $u(x) = \sin x$ 4. $u(x) = x^2$

5. $u(x) = \sec^2 x$

Exercises 4.2.2

1. $u(x) = \cosh x$ 2. $u(x) = x$

3. $u(x) = xe^x$ 4. $u(x) = x \sin x$

5. $u(x) = \sin x$ 6. $u(x) = x^3 - x^2 + x - 1$

7. $u(x) = \sin x$

Exercises 4.2.3

1. $u(x) = x \cos x$ 2. $u(x) = 1 - e^x$

3. $u(x) = \sin x - \cos x$ 4. $u(x) = 2x - 6x^2$

5. $u(x) = \frac{1}{2} \sin(2x)$

Exercises 4.3.1

1. $u(x) = x \cos x$ 2. $u(x) = 1 - e^x$

3. $u(x) = \sin x - \cos x$
4. $u(x) = 2x - 4x^2$
5. $u(x) = 1 - \sinh x$

Exercises 4.3.2

1. $u(x) = x + e^x$
2. $u(x) = 1 + \cos x$
3. $u(x) = 2e^x$
4. $u(x) = \sin x + \cos x$
5. $u(x) = 1 - \sin x$

Exercises 4.3.3

1. $u(x) = \cosh x$
2. $u(x) = \sin x$
3. $u(x) = \sinh x$
4. $u(x) = x + e^x$
5. $u(x) = \sin x + \cos x$
6. $u(x) = e^x$

Exercises 4.3.4

1. $u(x) = \frac{1}{2}\cos x + \frac{1}{2}\sin x + \frac{1}{2}e^x$
2. $u(x) = \sin x$
3. $u(x) = 1 + \sin x$
4. $u(x) = 1 + \cosh x$
5. $u(x) = 1 + 4x$
6. $u(x) = \frac{1}{4} + \sin x$
7. $u(x) = \frac{1}{4}e^x - \frac{3}{4}e^{-x} - \frac{1}{2}\cos x$

Exercises 5.2

1. $u(x) = 2\sqrt{x} + \frac{1}{\sqrt{x}}$
2. $u(x) = \sqrt{x}\left(\frac{4}{3}x - 1\right)$
3. $u(x) = \frac{1}{\pi\sqrt{x}}\left(1 + 2x + \frac{8}{3}x^2\right)$
4. $u(x) = x^{\frac{3}{2}}$
5. $u(x) = x$
6. $u(x) = \frac{1}{2}x^2$
7. $u(x) = \frac{16}{5\pi}x^{\frac{5}{2}}$
8. $u(x) = \frac{128}{35\pi}x^{\frac{7}{2}}$
9. $u(x) = \frac{2\sqrt{x}}{\pi}\left(1 + \frac{8}{5}x^2\right)$
10. $u(x) \approx \frac{2}{\pi}\sqrt{x}$

Exercises 5.3

1. $u(x) = \sqrt{x}$
2. $u(x) = x^{\frac{3}{2}}$
3. $u(x) = \frac{1}{2}$
4. $u(x) = \sqrt{x}$
5. $u(x) = x^{\frac{5}{2}}$
6. $u(x) = x^3$
7. $u(x) = 1 + x$
8. $u(x) = 1$
9. $u(x) = x^2$
10. $u(x) = \frac{2}{\pi}\sqrt{x} + \frac{15}{16}x^2$

Exercises 6.2.1

1. $u(x) = \dfrac{1 \pm \sqrt{1-2\lambda}}{\lambda}, \ \lambda \leq \dfrac{1}{2};$

   $\lambda = 0$ is a singular point, $\lambda = \frac{1}{2}$ is a bifurcation point

2. $u(x) = \dfrac{1 \pm \sqrt{-1+4\lambda}}{2\lambda}, \ \lambda \geq -\dfrac{1}{4};$

   $\lambda = 0$ is a singular point, $\lambda = \frac{-1}{4}$ is a bifurcation point

3. $u(x) = \dfrac{1 \pm \sqrt{1-2\lambda}}{\lambda}, \ \lambda \leq \dfrac{1}{2};$

   $\lambda = 0$ is a singular point, $\lambda = \frac{1}{2}$ is a bifurcation point

4. $u(x) = \dfrac{3 \pm \sqrt{9-12\lambda}}{2\lambda}, \ \lambda \leq \dfrac{3}{4};$

   $\lambda = 0$ is a singular point, $\lambda = \frac{3}{4}$ is a bifurcation point

5. $u(x) = \dfrac{2 \pm 2\sqrt{1-\lambda}}{\lambda}, \ \lambda \leq 1;$

   $\lambda = 0$ is a singular point, $\lambda = \frac{1}{2}$ is a bifurcation point

6. $u(x) = 2$

7. $u(x) = \sin x$
8. $u(x) = \cos x$
9. $u(x) = x,\ x + \dfrac{8}{3}$
10. $u(x) = x^2,\ x^2 + \dfrac{4}{3}$
11. $u(x) = x,\ x - 1$
12. $u(x) = x,\ x + \dfrac{1}{3}$

Exercises 6.2.2

1. $u(x) = 1 + \dfrac{\lambda}{2} + \dfrac{\lambda^2}{2} + \dfrac{5\lambda^3}{8} + \cdots$
2. $u(x) = 1 + \dfrac{\lambda}{4} + \dfrac{\lambda^2}{8} + \cdots$
3. $u(x) = 2\sin x$
4. $u(x) = 2\cos x$
5. $u(x) = \sec x$
6. $u(x) \approx 2x$
7. $u(x) = x^2$
8. $u(x) = x$
9. $u(x) = x$
10. $u(x) = x$
11. $u(x) = \sin x + \cos x$
12. $u(x) = \sinh x$
13. $u(x) = \cos x$
14. $u(x) = \sec x$

Exercises 6.3.1

1. $u(x) = x^2$
2. $u(x) = x^2$
3. $u(x) = 1 + x$
4. $u(x) = 1 + x^2$
5. $u(x) = x^2$
6. $u(x) = e^{-x}$
7. $u(x) = 1 + x$
8. $u(x) = \sin x$
9. $u(x) = \cos x$
10. $u(x) = e^x$

Exercises 6.3.2

1. $u(x) = 3x$
2. $u(x) = 2x$
3. $u(x) = \sin x$
4. $u(x) = x^2$

5. $u(x) = x + \frac{1}{20}x^5 + \frac{1}{720}x^9 + \cdots,$

6. $u(x) = 1 + \frac{1}{3}x^3 + \frac{1}{90}x^6 + \cdots$

7. $u(x) = 1 + \frac{1}{3}x^3 + \frac{1}{60}x^6 + \cdots$

8. $u(x) = x + \frac{1}{6}x^5 + \frac{1}{756}x^9 + \cdots$

9. $u(x) = 1 + 2x + \frac{5}{2}x^2 + \frac{1}{6}x^4 + \cdots$

10. $u(x) = 1 + x + x^2 + \frac{2}{3}x^3 + \frac{1}{6}x^4 + \cdots$

11. $u(x) = \sec x$

12. $u(x) = \tan x$

Bibliography

[1] G. Adomian, *Solving Frontier Problems of Physics: The decomposition method*, Kluwer, (**1994**).

[2] G. Adomian, *Nonlinear Stochastic Operator Equations*, Academic Press, San Diego,CA (**1986**).

[3] G.Adomian, A review of the decomposition method and some recent results for nonlinear equation, *Math. Comput. Modelling* **13***(7)*, 17–43,(1992).

[4] G. Adomian,and R. Rach, Noise terms in decomposition series solution, *Computers Math. Appl., Vol.* **24***(11)*, **61 – 64**, (**1992**).

[5] C. Baker, *The Numerical Treatment of Integral Equations*, Oxford University Press, London (**1977**).

[6] M. Bocher, *Integral Equations*,Cambridge University Press, London (**1974**).

[7] L. G. Chambers, *Integral Equations, A Short Course*, International Textbook Company, London (**1976**).

[8] Y. Cherruault, G. Saccomandi, and B. Some, New results for convergence of Adomian's method applied to integral equations, *Mathl. Comput. Modelling* **16***(2)*, 85 – **93**, (**1992**).

[9] Y. Cherruault, and G. Adomian, Decomposition methods: A new proof of convergence, *Mathl. Comput. Modelling* **18***(12)*, 103 – **106**, (**1993**).

[10] L. M. Delves, and J. Walsh, *Numerical Solution of Integral Equations*, Oxford University Press, London (**1974**).

[11] A. M. Golberg, *Solution Methods for Integral Equations: Theory and Applications*, Plenum Press, New York (1979).

[12] C. D. Green, *Integral Equations Methods*, Barnes and Noble, New York (1969).

[13] H. Hochstadt, *Integral Equations*, Wiley, New York (1973).

[14] A. J. Jerri, *Introduction to Integral Equations with Applications*, Marcel Dekker, New York (1985).

[15] R. P. Kanwal, *Linear Integral Equations, Theory and Technique*, Academic Press, New York (1971).

[16] W. V. Lovitt, *Linear Integral Equations*, Dover, New York (1950).

[17] G. Micula and P. Pavel, *Differential and Integral Equations through Practical Problems and Exercises*, Kluwer (1992).

[18] R. K. Miller, *Nonlinear Volterra Integral Equations*, W. A. Benjamin, Menlo Park, CA (1967).

[19] B.L. Moiseiwitsch, *Integral Equations*, Longman, London and New York (1977).

[20] F. G. Tricomi, *Integral Equations*, Interscience, New York (1957).

[21] H. J. TE Riele, Collocation methods for weakly singular second-kind Volterra integral equations with non-smooth solution, *IMA Journal of Numerical Analysis*, **2**, 437 – 449, (1982).

[22] V. Volterra, *Theory of Functionals and of Integral and Integro-Differential Equations*, Dover, New York (1959).

[23] A.M. Wazwaz, The decomposition method for approximate solution of the Goursat problem, *Applied Mathematics and Computation*, **69**, 299 – 311 (1995).

[24] A.M.Wazwaz and S. A. Khuri, Two Methods for Solving Integral Equations , *Applied Mathematics and Computation* , **77**, 79 – 89, (1996.

[25] A. M. Wazwaz and S. A. Khuri, A reliable technique for solving the weakly singular second-kind Volterra-type integral equations, *Applied Mathematics and Computation*, **80**, 287 – 299, (1996).

Index

Abel, Niels, 8
Abel's problem, 8, 141, 142
Abel's generalized integral equation, 8, 141, 146
Adomian, George, 33
Adomian decomposition method, 33, 68, 109, 163
Adomian polynomials, 164, 168, 178
Approximate solution, 13, 35
Bifurcation point, 160, 161
Boundary value problems, 26
Closed form, 13, 72
Convergence, 33, 67, 69, 112
Conversion,
  to Fredholm equations, 26, 118
  to Volterra equations, 20, 131
  to differential equations, 15, 82, 131
Decomposition method,
  for Fredholm equations, 33
  for Fredholm integro-differential equations, 109
  for Fredholm nonlinear equations, 163
  for Volterra equations, 68
  for Volterra integro-differential equations, 126
  for Volterra nonlinear equations, 177
  for weakly singular equations, 150
Direct computation method,
  for Fredholm equations, 43
  for Fredholm integro-differential equations, 105
  for Fredholm nonlinear equations, 159
Error function, 154, D.1
  complementary , 154, D.2
Fredholm alternative, 32
Fredholm integral equations, 31
  first kind, 3
  homogeneous, 59
  linear, 3
  nonhomogeneous, 31
  second kind, 4, 31
Fredholm integro-differential equations, 7, 105
Fredholm nonlinear equations, 158
Gamma function, 143, 146, D.3
Homogeneous differential equations, 83
Homogeneous integral equations, 6
Initial value problems, 2, 17, 82

Integral equation
- Abel's, 8, 141
- Fredholm, 3, 31
  - first kind, 3
  - second kind, 4, 31
- Homogeneous, 59
- Singular, 7, 8, 139
- Volterra, 3, 67
  - first kind, 4, 99
  - second kind, 4, 67

Integro-differential equations, 6, 103
Inverse operator, 109, 126, 143
Irrational functions, B.1-B.2
Kernels
- separable, 31
- singular, 7, 140, 150
- square integrable, 32

Laplace transforms, 140, 143, 146
- inverse, 143, 147

Leibnitz' rule, 16, 17, 102, 139
Modified decomposition method
- for Fredholm equations, 38
- for Fredholm integro-differential equations, 117
- for nonlinear Fredholm equations, 167
- for nonlinear Volterra equations, 179
- for Volterra equations, 73

Multiple integral, 20
Nonhomogeneous integral equations, 6, 9, 31, 67
Noise terms, 111, 120, 151
Nonlinear integral equations
- Fredholm, 158
- Volterra, 173

Numerical approximation
- Fredholm equations, 35
- Volterra equations, 72

Recurrent manner, 34, 39, 49, 74
Regularity condition, 32
Series representations, C.1-C.4
Series solution method
- for nonlinear Volterra equations, 173
- for Volterra equations, 77
- for Volterra integro-differential equations, 121

Singular integral equations, 139
Singular point, 160, 161
Solution
- exact form, 13, 46
- existence of , 13
- of an equation, 11
- series form , 13, 33, 46, 72
- uniqueness of , 13, 32
- truncated series, 35, 69

Successive approximations method, 48, 86
Successive substitutions method, 52, 91
Taylor series, 22, 71, 78
Volterra equations
- first kind, 3, 4, 67
- integro-differential, 104, 121
- linear, 67
- nonlinear, 173
- second kind, 4, 67
- weakly singular, 150

Weakly singular equations, 141
Zeroth approximation, 49, 60, 90
Zeroth component, 39